# The Alarm, Sensor & Security Circuit Cookbook

To Josh and Andy, and to electronic experimenters and hobbyists everywhere whose avocation of electronics, coupled with their creativity, energy, and determination, might help to solve our future energy and transportation problems so that our children can continue to breath clean air and drink pure water.

# The Alarm, Sensor & Security Circuit Cookbook

*Thomas Petruzzellis*

**TAB Books**
Imprint of McGraw-Hill

New York  San Francisco  Washington, D.C.  Auckland  Bogotá
Caracas  Lisbon  London  Madrid  Mexico City  Milan
Montreal  New Delhi  San Juan  Singapore
Sydney  Tokyo  Toronto

© 1994 by **TAB Books**.
TAB Books is a division of McGraw-Hill, Inc.

Printed in the United States of America. All rights reserved. The publisher takes no responsibility for the use of any of the materials or methods described in this book, nor for the products thereof.

pbk   2 3 4 5 6 7 8   9 10 11   DOC/DOC   9 9 8 7 6 5
hc      4 5 6 7 8 9 10 11 12   DOC/DOC   9 9 8 7 6

**Library of Congress Cataloging-in-Publication Data**

Petruzzellis, Thomas.
   The alarm, sensor, and security circuit cookbook / by Thomas Petruzzellis.
     p.  cm.
   Includes index.
   ISBN 0-8306-4314-1   ISBN 0-8306-4312-5 (pbk.)
   1. Detectors.  2. Electric alarms.  I. Title.
TK7870.P397   1993
681'.2—dc20                                              93-27562
                                                                                  CIP

Acquisitions editor: Roland S. Phelps
Editorial team:   Lori Flaherty, Managing Editor
                    Kenneth M. Bourne, Editor
                    Joann Woy, Indexer
Production team: Katherine G. Brown, Director
                     Rhonda E. Baker, Coding
                    Jan L. Fisher, Coding
                    Susan E. Hansford, Coding
                    Ollie Harmon, Coding
                    Lisa M. Mellott, Coding
                    Brenda M. Plasterer, Coding
                    Rose McFarland, Layout
                    Linda L. King, Proofreading
Design team: Jaclyn J. Boone, Designer
               Brian Allison, Associate Designer                          EL2
Cover design and illustration by Graphics Plus, Hanover, Pa.       4334

# Contents

Acknowledgments ix
Introduction xi

❖ **1 Sensors and detection circuits** 1
High-gain amplifier  *1*
Touch switch  *5*
Static-electricity detector  *7*
Electroscope  *8*
Light/dark switch  *9*
Temperature sensors  *13*
Hall-effect metal/ magnetic sensors  *19*
ac-magnetic field detector  *25*
Bridge sensors  *27*
Maxwell bridge  *31*
Pyroelectric detector  *33*
Pressure sensor  *36*
Toxic-gas sensor  *40*
Humidity sensor  *42*
Fiberoptic sensing  *44*
Optical rotary encoder  *47*
Film strip position sensor  *50*
Tiltmeter  *52*
Earth-movement sensor  *59*

❖ **2 New film sensor technology** 63
Piezo film sensors  *63*
Force and position-sensing resistor  *72*

❖ 3  **New sensors, ICs, and gas-sensing technology**    **79**
　　Piezo accelerometer   *79*
　　Optical transceiver   *81*
　　Proximity detector   *83*
　　Smoke detector   *85*
　　Fluid detector   *87*
　　Over/under current detector   *89*
　　Tachometer/speed detector   *91*
　　Position-sensitive detector   *94*
　　Twilight sensor   *97*
　　Multiplexed Hall-effect sensor   *98*
　　Videophone   *103*
　　Gas-sensing technology   *105*

❖ 4  **Computer interfacing**    **125**
　　Joystick interfaces   *125*
　　Serial interface   *127*
　　Mouse and trackball interface   *128*
　　Analog-to-digital interface   *129*

❖ 5  **Describing and surveying alarm systems**    **133**
　　Window foil   *133*
　　Magnetic switches   *134*
　　Holdup switch   *135*
　　Sensor mats   *136*
　　Motion sensors   *136*
　　Smoke and fire sensors   *145*
　　Cameras   *147*

❖ 6  **Alarm-system design philosophy**    **151**
　　Doors, keys, and locks   *151*
　　Designing your alarm   *153*
　　Installation and wiring tips   *154*
　　Alarm history   *156*
　　High-tech control boxes   *159*
　　Sirens, strobe lights, and phone dialers   *160*
　　Wireless alarm systems   *160*
　　Fire reporting   *162*
　　Digital telephone dialers   *163*
　　Lighting for crime prevention   *164*
　　Detecting bugs   *166*

### ❖ 7 Alarm circuits and systems — 169

Basic latching alarm  *169*
Remote sensing  *171*
Window/door alarm  *173*
Alarm system with location display  *175*
Multipurpose alarm  *178*
Auto burglar alarm  *181*
Bar-graph auto voltmeter  *183*
Auto immobilizer  *184*
Digital antitheft auto lock  *186*
Power-line fault detector  *188*
Automatic emergency-lighting system  *190*
Adjustable-rate siren  *191*
Alarm strobe flasher  *193*
Telephone line monitor  *195*
Automatic telephone recorder  *196*
rf sniffer  *198*
Home guard  *201*
Optical motion detector  *205*
Capacitive proximity sensor  *207*
Microwave motion detector  *209*
Carrier-current control systems  *211*
Ultrasonic-sensor system  *217*
High-performance alarm  *220*

### ❖ 8 Unique, high-tech security projects — 223

Piezo vibration alarm  *223*
Camp alarm  *226*
Pyroelectric sensor  *230*
Chimney alarm  *235*
DTMF alarm system  *240*
Portable alarm  *247*
Storm warn  *258*
Video sentry  *264*

### Appendix Suppliers — 275

### Index — 279

# Acknowledgments

A BRIEF THANK YOU IS MADE TO THE FOLLOWING SEMICONDUCTOR manufacturers for the circuit diagrams used in this book. Credits for circuits shown use two- or three-letter abbreviations placed near schematic diagrams.

Cherry Semiconductor Corp. (CS)
2000 South County Trail
East Greenwich, RI 02818
(401) 885-3600

Eltec Instruments, Inc. (EI)
350 Fentress Boulevard
Daytona Beach, FL 32114
(800) 874-7780

Hamamatsu Corp. (HC)
360 Foothill Road
P.O. Box 6910
Bridgewater, NJ 08807
(201) 231-0960

National Semiconductor (NS)
2900 Semiconductor Drive
P.O. Box 58090
Santa Clara, CA 95052
(408) 721-5000

Penwalt Corp. (PW)
P.O. Box 799
Valley Forge, PA 19482
(215) 666-3500

PMC Electronics (PMC)
P.O. Box 11148
Marina del Ray, CA 90292

Sensym, Inc (SM)
1244 Reamwood Avenue
Sunnyvale, CA 94089
(408) 744-1500

Allegro Corp. (AMS)
formerly Sprague
P.O. Box 15036
Worcester, MA 01615
(508) 853-5000

Bourns Inc. (BR)
1200 Columbia Avenue
Riverside, CA 92507

Interlink Corp.
110 Mark Avenue
Carpinteria, CA 93013

Gernsback Publishing Co. (GB)
500 B Bi-County Boulevard
Farmingdale, NY 11735

Fredericks Co. (FR)
Philmont Avenue & Anne Street
Huntington Valley, PA 19006

Figaro Engineering Co. (FI)
1000 Skokie Blvd. Rm 357
Wilmette, IL 60091

Forrest Mims, III (FM)
433 Twin Oaks Rd.
Seguin, TX 78155

Texas Instruments Inc. (TI)
P.O. Box 655503
Dallas, TX 75265

Ramsey Electronics (RE)
793 Canning Parkway
Victor, NY 14564

LSI Computer Systems
1235 Walt Whitman Road
Melville, NY 11747

Optoelectronics (OE)
5821 NE 14th Avenue
Ft. Lauderdale, FL 33334

# Introduction

MOST PHYSICAL PHENOMENA CAN BE DETECTED BY SENSORS, monitored by amplifiers and trigger circuits, and then presented by meters, bells, sirens, chart recorders, or personal computers. Measurement and protection systems utilize sensors and detectors that can be used to detect light, temperature, pressure, speed, vibration, proximity, infrared, metal/magnetism, acceleration, and toxic gases.

One of the aims of this book is to present the many types of sensors that can be used in measurement and protection circuits in a "cookbook" of ideas and circuits that can be called upon when a particular problem or application arises. This book should appeal to engineers, technicians, alarm installers, and hobbyists.

Many new ideas and integrated circuits are introduced, so you can become familiar with the latest sensors, detection circuits, and integrated circuits available. The scope of this book includes both sensing and measurement devices, as well as stand-alone alarm circuits. Many of the sensors shown can be wired together to form more complex protection or alarm circuits.

The first chapter begins with the high-gain amplifier and how it can be used in a multitude of sleuthing applications, including detecting light, sound, motion, radiation, magnetism, and rf energy. You might not have realized just how many phenomena can be sensed with the lowly amplifier. Following the high-gain amplifier, many types of sensors are shown, such as a static-electricity detector, light and heat detectors, temperature sensors, and metal and magnetic sensors. The measurement-bridge circuit is described next and it is shown in a variety of different configurations. An ac Maxwell bridge follows. It can measure unknown capacitance or inductance. It is shown as an automobile metal sensor, which can detect cars passing over a driveway. Hall-effect sensors are presented next; and they can detect metal, magnetism, speed, pressure, and current flow. The next detectors include a pyroelectric or infrared body-heat sensor, pressure sensors, a toxic-gas detector, optical encoders, and tiltmeters.

Chapter 2 presents the revolutionary piezoelectric film. This new material can be used in a spectacular array of sensing applications. The piezo film is now used in many types of sensors including vibration switches, magnetic switches, infrared sensors, fluid sensors, microphones, hydrophones and accelerometers, and the list is growing. We will suggest how you may obtain a sample of this amazing material. Chapter 2 also introduces the new force-sensing resistance sensors.

Chapter 3 introduces a number of new integrated circuits that can be used to build low-cost, minimum-component sensing systems, such as a proximity sensor, speed detector, smoke detector, and precision position detector. We also present a new video transceiver chip, which can be configured into a high-resolution videophone or the unique video sentry described in Chapter 8. Chapter 3 also discusses gas sensors and recent trends in gas-sensing technology.

Chapter 4 is devoted to computer interfacing. A number of low-cost methods are described to help you interface sensing and measurement circuits to the personal computer so you can collect, store, and display your measurement data.

Chapter 5 surveys some of the most often used alarm-system sensors. An overall view of each sensor is presented, and strengths and weaknesses are discussed. Recommendations are made for the most suitable use for each device.

Chapter 6 is a short course on the philosophy of alarm-system design. Useful tips are discussed, as well as the pitfalls of alarm systems. Thoughts on how burglars think and how to outsmart the common thief are presented.

Chapter 7 includes diagrams of alarm systems that can protect your home or office. Shown first is the basic latching alarm, which is the heart of most alarm systems. Next, is a remote sensing system that can be used to take measurements of light, temperature, and speed and send the data over a wire or radio-frequency (rf) link to a remote monitoring site. A low-cost window/door alarm is shown, which can be configured to protect most doors and windows. Next is a unique security system that displays each alarm location and status. It can call the local police department. Next is a multipurpose, dual-channel alarm system that can monitor both fire and alarm conditions. Last, a number of circuits, including low-cost automobile alarms, emergency lighting, strobes, sirens, phone circuits, and motion sensors are presented.

In the last chapter, a number of novel, high-tech detection and alarm projects are covered. Each circuit includes a circuit board layout to aid in constructing the particular project. The first circuit is a sensitive piezoelectric vibration sensor, which can be implemented as a complete stand-alone travel alarm or wired with other sensors for a more complex alarm system. The next project is a self-contained camping alarm system, followed by a pyroelectric infrared body-heat detector. The pyroelectric sensor is one of the most sensitive and trouble-free detectors available. It can sense humans or large animals up to 50 feet away.

A unique high-chimney alarm is the next project. The chimney alarm senses an overheating chimney, triggers the alarm, calls the fire department, and extinguishes the chimney fire, all simultaneously. The tone-identification alarm is a useful project that identifies a particular location which has been activated by sending a Touch Tone signal from one of the trigger modules to the decoder/display unit. The beauty of this system is that it can be used over either a hard-wire or rf link.

The portable alarm is one of my favorite projects. It is a wireless infrared system that alerts your friends or neighbors when an intruder has entered your home or cabin. A pyroelectric sensor that can detect humans up to 50 feet is used to trigger a transmitter that sends an alerting tone to an FM receiver or scanner for 20 seconds. After the 20-second time period, a sensitive microphone is connected to the transmitter, allowing your neighbor or friend to "listen in" to your home for up to five minutes, at which time the system resets. During this five-minute period, your neighbor could investigate or call the police, if necessary.

Next is the storm-warn project. It can disconnect computers or antennas during an electrical storm.

The last project is the video sentry, a sophisticated audio/video security/surveillance system that permits you to monitor both audio and video from a distant location where the video sentry is installed, using the public telephone network. You can monitor your office while you are away or keep tabs on your babysitter or old folks as well. This modern-day infinity transmitter operates over any geographic distance and it is simple to operate. Simply dial the phone number where the video sentry was installed, press a Touch Tone function key on your phone, and the video sentry at the remote location will instantly and automatically answer the phone line without even ringing the phone. The video sentry also allows you to control remotely a number of devices such as bells, sirens, tape recorders, lamps, and home appliances. The video sentry is possible because of a new videophone chip, which sends high-resolution still pictures over a twisted pair in less than 12 seconds. The PMC videophone chip produces the best picture of any videophone offered to date.

# ❖ 1
# Sensors and detection circuits

SENSORS ARE THE WINDOWS TO THE WORLD! THE HUMAN SENSES are limited to a narrow range of audio and video frequencies. For us to detect the broad range of physical phenomena all around us, we often rely on the magic of electronics. Electronic sensors provide us with the means to augment the human senses to detect pressure, motion, radiation, infrared, gases, etc.

## High-gain amplifier

A simple but highly effective means to sense or monitor physical phenomena can be accomplished by using a high-gain amplifier, as shown in Figs. 1-1 and 1-2. You can easily become a real sleuth using the sensitive amplifier. This lowly device can monitor all sorts of things. Everyone is aware that connecting a microphone to an amplifier permits you to listen to nearby sounds and placing a microphone at the focus point of a parabola allows you to hear distant sounds. But did you know you could connect a small crystal earphone to the input of a high-gain amplifier and by epoxying the earphone to a nail pounded halfway through a wall you would have an extremely powerful listening device, one that could listen through walls?

Have you ever thought about connecting a ceramic phono cartridge to a high-gain amplifier? You can epoxy an 8 to 10 inch brass rod to a phono cartridge to create a vibration monitor, which you could use to listen for a bad bearing in a motor. Try winding a small 100-turn coil of 28-gauge enameled wire around a ferrite or iron core. Connect the coil to your high-gain amplifier and you can "listen in" to ac wiring inside walls to help locate hidden wiring (see Fig. 1-3). If you placed the same coil near a telephone,

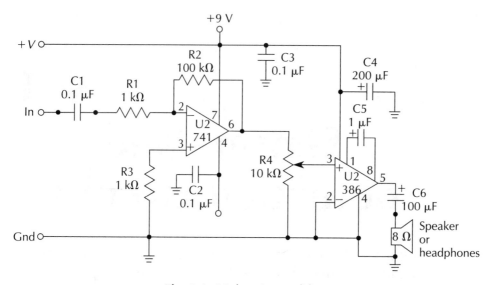

**Fig. 1-1** High-gain amplifier.

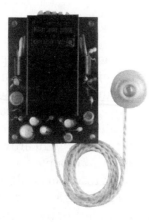

**Fig. 1-2** Microphone and high-gain amplifier.

you would have a telephone amplifier that would amplify a long-distance phone call or perhaps provide a remote ringer in another room. If you placed a magnet at a 30 to 40° angle as shown in Fig. 1-4, you could listen for hidden nails in a plaster wall.

Listening to nature's sounds can be accomplished quite easily by connecting six to eight turns of 26-gauge wire wound on a 3 × 5-foot loop placed outside. You can listen to a "dawn chorus" or lightning flashes, atmospheric whistlers, even auroras. The basic high-gain amplifier can also be used to detect radio frequency (rf) energy from radio or television transmitters, so you

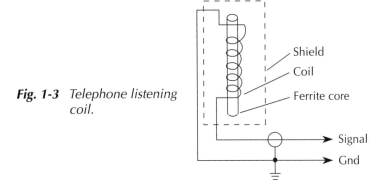

**Fig. 1-3** *Telephone listening coil.*

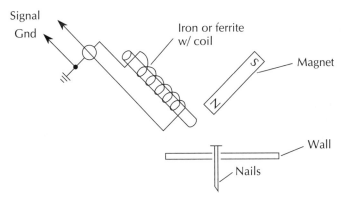

**Fig. 1-4** *Metal detector.*

now have a "bug" detector or a field-strength meter. Connect a single loop coil antenna to a high-frequency diode to the input of your high-gain amplifier. As shown in Fig. 1-5, you can sniff rf energy. By connecting a hydrophone (a ceramic underwater microphone) to the high-gain amplifier, you can construct a fish finder, a pool splash detector, or a marine engine detector.

If you connect a solar cell as a sensor to the high-gain amplifier as shown in Fig. 1-6, you can use the silicon solar cell to sense the speed of a propeller or any rotating object by shining a light on the rotating object and placing the solar cell in view of the rotating object. A solar-cell detector can be used to detect lights in the night sky by placing a telescope or lens in front of the solar cell. You can also "listen" to airplane strobe lights or perhaps you could construct a moonlight detector to steer your telescope.

**4** Sensors and Detection Circuits

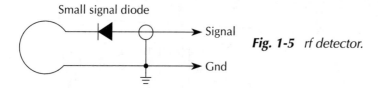

**Fig. 1-5** *rf detector.*

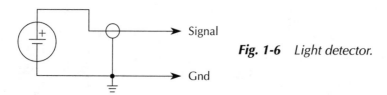

**Fig. 1-6** *Light detector.*

You can even measure radiation with the aid of a high-gain amplifier. One of the most sensitive forms of radiation detectors is the scintillator. When radiation strikes a crystal, it scintillates, emitting a small amount of light. That light can be detected by a silicon solar cell, as shown in Fig. 1-7. A simple detector can be constructed by using two microscope slides, a solar cell, and some zinc sulfide, which is easily obtainable. Mix the zinc sulfide into a slurry, using ordinary tap water. Then place the slurry on one of the microscope slides. When the slide is dry, place the other slide over the coating and tape the two slides together at the edges. Then position the scintillator in front of the silicon solar cell and place the detector in a dark enclosure. Allow 10 minutes to recover from the ambient light. This detector arrangement allows you to detect gamma rays by listening to clicks in a headphone connected to your high-gain amplifier.

The next time you need to sniff out a problem, don't forget your trusty friend, the high-gain amplifier.

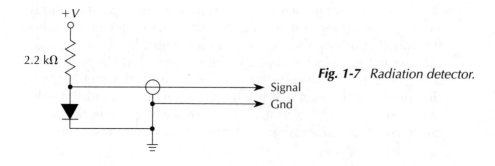

**Fig. 1-7** *Radiation detector.*

## High-gain amplifier parts list

| Quantity | Part | Description |
|---|---|---|
| 2 | R1, R3 | 1-kΩ, ¼-W resistor |
| 1 | R2 | 100-kΩ, ¼-W resistor |
| 1 | R4 | 10-kΩ potentiometer, ½W |
| 3 | C1, C2, C3 | 0.1-µF, 25-V capacitor |
| 1 | C5 | 1-µF, 25-V capacitor |
| 1 | C6 | 100-µF, 25-V electrolytic capacitor |
| 1 | C4 | 200-µF, 25-V electrolytic capacitor |
| 1 | U1 | UA741 op amp |
| 1 | U2 | LM386 audio amplifier |
| 1 | SP | 8-Ω speaker |

## Touch switch

A touch switch is a useful circuit that can be used to detect humans or protect small objects, such as antiques. It can be used to turn on a lamp or as an annunciator to sound a buzzer when someone comes near a door or table. The touch switch, or capacity switch, can also be used to start a moving display sign. A touch switch is shown in Fig. 1-8, and it can be activated by touching a small metal plate connected to pin 2 of the 555 timer chip. Once triggered, the load remains on until reset. A low logic level applied to pin 4 resets the circuit. The output is on pin 3, which is used to drive an LED.

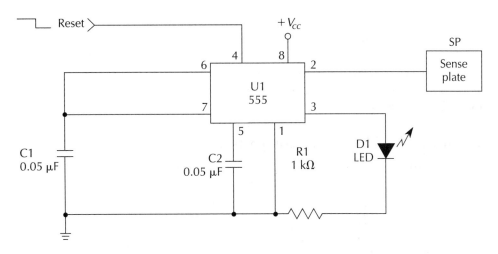

*Fig. 1-8* Touch switch—manual reset.

## 6 Sensors and Detection Circuits

Another variation of the touch switch is depicted in Fig. 1-9. This touch switch also uses the ubiquitous 555 chip. The circuit is configured as a monostable multivibrator. The load remains on for a time period determined by the R1/C1 combination. After the time period elapses, the circuit turns off until triggered again. The sense plate is connected to a capacitor placed in series with pin 2 of the IC timer to increase the charge accumulation.

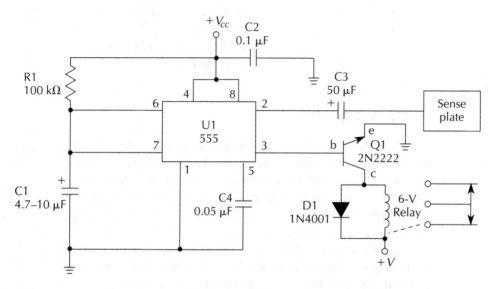

**Fig. 1-9** Touch switch.

The touch switch relies on the "stray capacitance effect" of a human body from the sense plate to a lower potential, i.e., ground. By completing a path to ground through the human body, the switch magically appears to turn on a light or external load. Always power your touch switch either with batteries or with a power supply that uses a transformer to ensure you are not in the direct path to a 110-$V_{ac}$ line.

### Touch switch with manual reset parts list

| Quantity | Part | Description |
|---|---|---|
| 1 | R1 | 1-kΩ, ¼-W resistor |
| 2 | C1, C2 | 0.5-µF, 25-V capacitor (disk) |
| 1 | D1 | Red LED |
| 1 | U1 | 555 timer IC |
| 1 | S1 | Sense-plate copper circuit board |

**Touch switch parts list**

| Quantity | Part | Description |
|---|---|---|
| 1 | R1 | 100-kΩ, ¼-W resistor |
| 1 | C1 | 4.7–10-µF, 25-V electrolytic capacitor |
| 1 | C2 | 0.1-µF, 25-V capacitor |
| 1 | C4 | 0.05-µF, 25-V capacitor (disk) |
| 1 | C3 | 50-µF, 25-V electrolytic capacitor |
| 1 | D1 | 1N4001 silicon diode |
| 1 | Q1 | 2N2222 pnp transistor |
| 1 | U1 | 555 IC timer |
| 1 | Ry-1 | 6-V SPST relay |
| 1 | S1 | Sense-plate copper circuit board |

# Static-electricity detector

The static-electricity detector shown in Fig. 1-10 is a simple tester designed to detect nearby static-electricity fields. You can easily demonstrate a static field by walking across a carpet and then touching the sensor probe. When the detector is placed next to a television screen or computer monitor, it is activated by the high voltage that accelerates electrons in the picture tube. A cellophane tape roll also generates a static charge. Place the probe wire near where the tape comes off the roll. Then pull the tape through the dispenser and the meter will move.

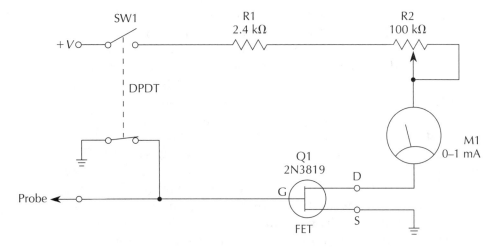

**Fig. 1-10** *Static-electricity detector.*

A 2N3819 field-effect transistor is used as the static field sensor. A shore wire or small telescoping radio antenna is connected to the gate of the FET. The source lead is connected to ground and the drain lead is connected to a 0–1-mA meter. The remaining meter lead is coupled to a 3.3-k$\Omega$ resistor, which is fed to the positive post of a 9-V transistor-radio battery. Note that the FET can be easily damaged with a high static field while it is being handled. The FET leads should be shorted together as it is soldered in place and a grounded soldering pencil should be used. A grounded wrist band is also recommended. The static sensor would make a great addition to any static-electricity science-fair project, or could be a handy sensor on your test bench.

**Static-electricity detector parts list**

| Quantity | Part | Description |
| --- | --- | --- |
| 1 | R1 | 2.4-k$\Omega$, ¼-W resistor |
| 1 | R2 | 100-k$\Omega$, ¼-W resistor |
| 1 | Q1 | 2N3819 FET |
| 1 | M | 0–1-mA meter |
| 1 | SW-1 | DPST toggle switch |
| 1 | ANT | Whip antenna or wire |

# Electroscope

The electroscope pictured in Fig. 1-11 can be used to display static energy charges from sources such as TV sets, electrostatic generators, carpet cruising, and hair combing. The electroscope is the sophisticated cousin of the static-electricity detector shown in Fig. 1-9. The electroscope would make an excellent science-fair project or addition to your electronics bench.

The heart of the electroscope circuit is the two FETs, Q1 and Q2, connected in a balanced bridge configuration. The gate of Q1 is connected to the wire pick-up antenna via a 1.5-$\Omega$ resistor, and the gate of Q2 is tied to the circuit's common ground through the other 1.5-$\Omega$ resistor. This type of bridge circuit offers excellent temperature stability. Q1 operates in an open-gate configuration. The 500-$\Omega$ potentiometer balances the null bridge circuit. The 5-k$\Omega$ potentiometer and capacitors C1 and C2 help reduce stray 60-Hz pickup and increase the stability of the circuit. The 1-mA meter connected between the drain pins of Q1 and Q2 indicates an electrostatic field. The electroscope requires little current consumption, and therefore, it can be operated from a 9-V transistor-radio battery.

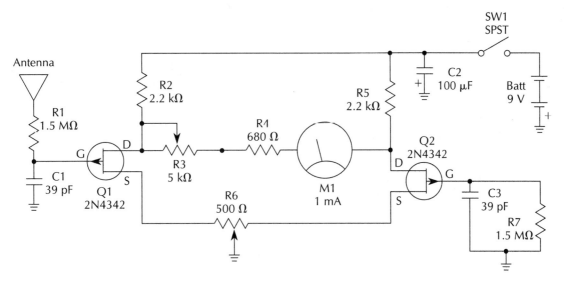

*Fig. 1-11* Electroscope.

**Electroscope parts list**

| Quantity | Part | Description |
|---|---|---|
| 2 | R1, R7 | 1.5-Ω, ¼-W resistor |
| 2 | R2, R5 | 2.2-kΩ, ¼-W resistor |
| 1 | R3 | 5-kΩ, ¼–12-Watt potentiometer |
| 1 | R4 | 680-Ω, ¼-W resistor |
| 1 | R6 | 500-Ω, ¼-Watt potentiometer |
| 2 | C1, C3 | 39-pF, 25-V capacitor |
| 1 | C2 | 100-µF, 25-V electrolytic capacitor |
| 2 | Q1, Q2 | 2N4342 FET |
| 1 | M | 0–1-mA panel meter |
| 1 | SW-1 | SPST toggle switch |
| 1 | ANT | Telescoping whip antenna |
| 1 | BATT | 9-V transistor-radio battery |

# Light/dark switch

The light/dark switch can be used in many sensing or alarm circuits. The light/dark switch shown in Fig. 1-12 can detect an intruder passing through a light beam or a person moving through a normal ambient-light room by keeping the lighting constant. The light/dark detector can also be used as an annunciator to inform you of an approaching customer in a retail store. The circuit can

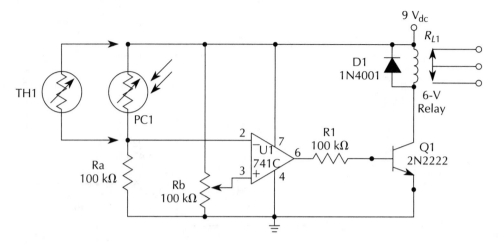

**Fig. 1-12**  *Light/dark detector.*

also detect your automobile headlamps as you approach your home and turn on your home's lights. The light/dark sensor can wake you at dawn or start your coffee pot in the morning. You can use the light/dark sensor as the heart of a laser tag game for your children. Various types of light sensors are shown in Fig. 1-13.

**Fig. 1-13**  *Light detectors.*

The light/dark switch uses an LM741 op amp as a comparator. You can substitute an LM339 or any general purpose op-amp pin in this design. The adjustable-threshold detector is controlled by Rb, a 100-kΩ potentiometer. Rb sets the threshold value between the voltage divider of PC1 and Ra, a 100-kΩ resistor. When the light intensity at PC1 is increased, its resistance decreases. This increases the voltage on pin 2 on the op amp's inverting-input pin. When the reference voltage at pin 3 has been exceeded by the input voltage on pin 2, the comparator will present an output on pin 6. The output drives Q1, a 2N2222 transistor, which can be used to drive a small relay. To make your detector more efficient, consider using a black plastic or cardboard tube with the sensor mounted at one end of the tube (see Fig. 1-14). The tube reduces the field of view and helps to prevent unwanted ambient light from reaching the detector. To use the sensor with ambient light as the input source, use a small light tube. To create a "beam type detector" system, use a long light tube. To construct a long-range detector, place a lens in front of the detector inside the light tube. The focal distance is determined by the lens you select. The same light/dark sensor circuit can also sense temperature by using a negative-temperature coefficient thermistor. A room-temperature resistance of 20–50 kΩ is needed for the thermistor. The accuracy of your temperature switch is determined by your selection of components and the method of calibration.

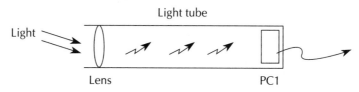

**Fig. 1-14** *Light tube.*

A light-detector circuit that detects specific levels of light is shown in Fig. 1-15. When the light level goes above or below the desired set values, the window comparator circuit activates a low-current relay, which can be connected to an alarm buzzer. The window comparator circuit uses two sections of an LM339 op amp to act as a specific-level light detector. R1 and R3 are the high/low-value potentiometers. The op-amp outputs are wired

## 12 Sensors and Detection Circuits

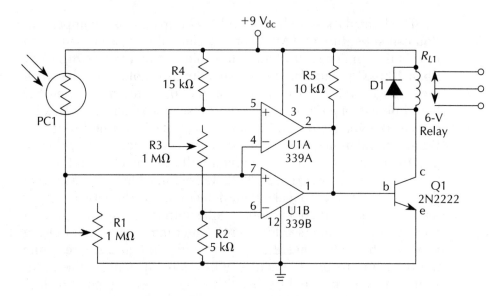

**Fig. 1-15** Light-level detector.

together and drive a 2N2222 transistor, which can drive a low-current relay. The comparator circuit can be powered by a 9 to 12-V power source or a common 9-V battery, if desired. Simply adjust R1 and R3 to the set-point values so that the relay pulls in when the light level at PC1 is above or below the desired value.

## Light/dark detector parts list

| Quantity | Part | Description |
|---|---|---|
| 1 | TH1 | Thermistor |
| 1 | PC1 | Cadmium photoresistive cell |
| 3 | RA, RB, R1 | 100-kΩ, ¼-W resistor |
| 1 | D1 | 1N4001 silicon diode |
| 1 | Q1 | 2N2222 pnp transistor |
| 1 | U1 | UA741C op amp |
| 1 | RY-1 | 6-V SPST relay |

## Light-level detector parts list

| Quantity | Part | Description |
|---|---|---|
| 2 | R1, R3 | 1-MΩ potentiometer (trim) |
| 1 | R2 | 5-kΩ, ¼-W resistor |
| 1 | R5 | 10-kΩ, ¼-W resistor |
| 1 | R4 | 15-kΩ, ¼-W resistor |

| | | |
|---|---|---|
| 1 | PC1 | Cadmium photoresistive cell |
| 1 | D1 | 1N4001 silicon diode |
| 1 | Q1 | 2N2222 pnp transistor |
| 1 | U1 | LM339A comparator IC |
| 1 | RY-1 | 6-V SPST relay |

## Temperature sensors

The need to measure temperature often arises and temperature sensing has become easy with many new precision temperature sensors. Figures 1-16 and 1-17 illustrate a number of thermistors and thermocouples. The LM34 is a precision Fahrenheit temperature-sensing IC and the LM35 is a precision centigrade temperature sensor. Both sensors require no external calibration and have accuracies of better than ±1.2° at room temperature. The LM34/35 draws only 70 µA and has a low output impedance. Both sensors operate from 5 to 30 $V_{dc}$.

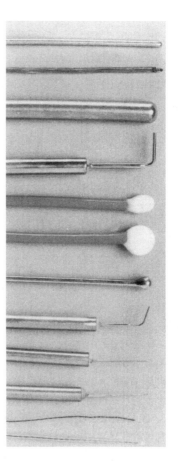

***Fig. 1-16*** *Thermistors.*

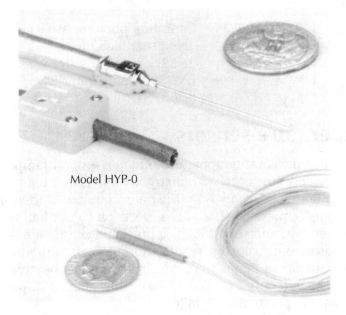

**Fig. 1-17** *Thermocouples.*

An analog temperature sensor and display can be constructed simply by connecting the sensor to a 100-μA meter via a potentiometer, as shown in Fig. 1-18. An extended-range room temperature meter is shown in Fig. 1-19. The LM34/35 is easy to use in all types of temperature-sensing applications.

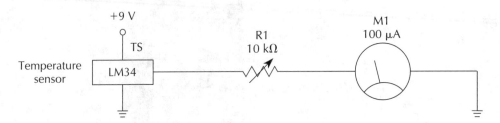

**Fig. 1-18** *Temperature sensor.*

Long-distance or remote monitoring of the sensors often is difficult due to large capacitances across the signal leads. Figure 1-20 illustrates two recommended circuits to solve this problem. The first diagram uses a 2-kΩ resistor to isolate the load. The second diagram illustrates the load separation via an RC damper, which uses a 75-Ω resistor and a 10-μF capacitor across the sensor output.

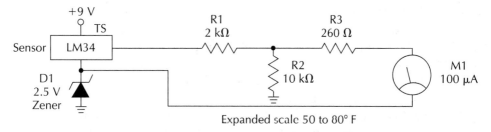

**Fig. 1-19** *Extended-range temperature sensor.*

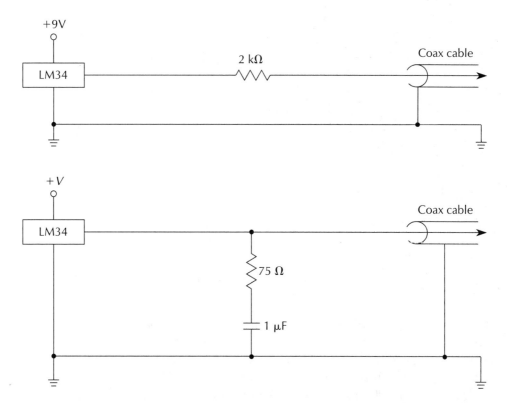

**Fig. 1-20** *Temperature-sensor interfacing.*

One common concern with temperature sensors is the thermal coupling from the sensor and the surface to be sensed, and the thermal differences between the metal leads and the sensor's plastic package. To solve these problems, try placing beads of epoxy over the leads, covering them completely. Use epoxy with good thermal-transfer characteristics to bond the sensor to the heat-source surface.

Figure 1-21 depicts how the LM34/35 temperature sensors could be connected to a personal computer via a parallel data input. An LM34 is connected to a signal conditioner, such as a 75-Ω resistor and 1-μF capacitor. An op amp sets up a voltage reference to the A/D converter, an ADC0804 8-bit A/D chip. A parallel data stream and the control signals $\overline{CS}$, $\overline{RD}$, $\overline{WR}$, and $\overline{INTR}$ are sent to the computer.

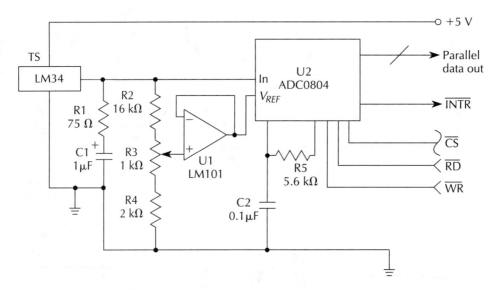

**Fig. 1-21** Computer interface for temperature sensor.

Next, a high-precision temperature sensor, an LM135, operates as a zener diode. The LM135 has a breakdown voltage directly proportional to absolute temperature 10 mV/°K. The device has a 1-Ω impedance and operates from 400 μA to 5 mA with less than 1°C error over a 100°C range. The low output impedance and linear output make this detector a good choice for temperature-sensing applications (see Fig. 1-22).

Calibration of the LM135 requires a 10-kΩ potentiometer placed across the sensor, as shown in Fig. 1-19. The calibration voltage should be 2.98 V at 25°C.

The last example of a temperature sensor circuit uses the LM335 precision temperature sensor, which is coupled to an LM311 op amp via pin 3. A voltage reference and adjustment is

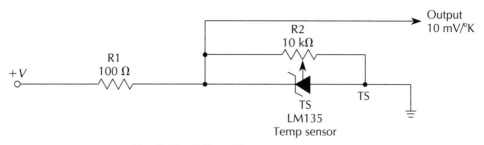

**Fig. 1-22** *Adjustable temperature sensor.*

set up using an LM329 electronic zener diode and a 10-kΩ potentiometer connected to pin 2 of the IC (see Fig. 1-23). The output of the LM311 is then coupled to an LM395 power npn transistor, which can drive a heating element or motor. Further application assistance can be obtained by calling National Semiconductor at 1-800-272-9959.

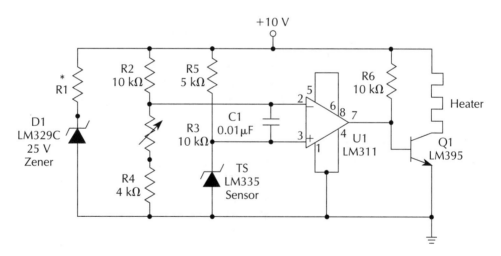

**Fig. 1-23** *Temperature-controlled heater.*

## Temperature sensor parts list

| Quantity | Part | Description |
|---|---|---|
| 1 | TS | LM34 temperature sensor (National Semiconductor) |
| 1 | R1 | 10-kΩ, ¼-W potentiometer |
| 1 | M | 100-μA panel meter |

## Extended-range temperture sensor parts list

| Quantity | Part | Description |
|---|---|---|
| 1 | TS | LM34 temperature sensor (National Semiconductor) |
| 1 | R1 | 2-kΩ, ¼-W resistor |
| 1 | R2 | 10-kΩ, ¼-W resistor |
| 1 | R3 | 260-kΩ, ¼-W resistor |
| 1 | D1 | 2.5-V sensor diode |
| 1 | M | 0–100-mA panel meter |

## Computer interface for temperature sensor parts list

| Quantity | Part | Description |
|---|---|---|
| 1 | TS | LM34 temperature sensor |
| 1 | R1 | 75-Ω, ¼-W resistor |
| 1 | R2 | 16-kΩ, ¼-W resistor |
| 1 | R3 | 1-kΩ, ¼–½-W potentiometer |
| 1 | R4 | 2-kΩ, ¼-W resistor |
| 1 | R5 | 5.6-kΩ, ¼-W resistor |
| 1 | C2 | 0.1-µF, 25-V capacitor |
| 1 | C1 | 1-µF, 25-V capacitor |
| 1 | U1 | LM101 op amp |
| 1 | U2 | ADC0804 A/D converter (National Semiconductor) |

## Adjustable temperature sensor parts list

| Quantity | Part | Description |
|---|---|---|
| 1 | R1 | 100-Ω, ½-W resistor |
| 1 | R2 | 10-kΩ, ½-W potentiometer |
| 1 | TS | LM135 temperature sensor (National Semiconductor) |

## Temperature-controlled heater parts list

| Quantity | Part | Description |
|---|---|---|
| 1 | TS | LM335 temperature sensor (National Semiconductor) |
| 1 | R1 | 100–500-Ω, ½-W resistor |
| 1 | R2 | 10-kΩ, ½-W resistor |
| 1 | R3 | 10-kΩ, ½-W potentiometer |
| 1 | R4 | 4-kΩ, ½-W resistor |
| 1 | R5 | 5-kΩ, ½-W resistor |

| | | |
|---|---|---|
| 1 | C1 | 0.01-µF, 25-V capacitor (disk) |
| 1 | D1 | LM329C zener diode (National Semiconductor) |
| 1 | Q1 | LM395 transistor (National Semiconductor) |
| 1 | H | Heater coil |

## Hall-effect metal/magnetic sensors

The Hall-effect sensor can detect the pressure or absence of magnets or metal surfaces and can solve a number of measurement and alarm problems. We will illustrate some of these applications. Three different types of Hall-effect sensor are presented. Hall-effect sensors can be used for tilt switches, counters, proximity detectors, metal detectors, pressure sensors, current sensors, and gaussmeters.

A basic low-cost Hall-effect sensor is the Sprague UGN3020T, shown in Fig. 1-24. This sensor includes a Hall cell, voltage regulator, signal amplifier, and Schmitt trigger. The circuit shown can protect instruments, antiques, or windows/doors by placing a small magnet on the item you wish to protect. The sensor is then placed in close proximity to the magnet. When the magnet is moved past the sensor, an output activates a relay. The sensor drives a 2N5812 transistor for an on-off type output. A DPST relay could be used to sound a local alarm, or the circuit could be con-

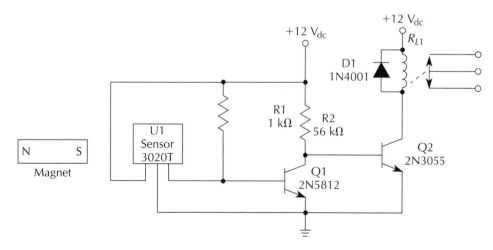

***Fig. 1-24*** *Hall-effect switch.*

nected to a central alarm panel. The same basic Hall sensor could be used for counting moving objects or as a wind-speed indicator.

The diagram in Fig. 1-25 depicts an ac Hall switch that can control 115-$V_{ac}$ loads such as motors and lamps, upon detecting the presence or absence of a magnet. The Sprague UGN3501T Hall sensor is more flexible because it provides an analog output. The 350IT Hall sensor is a single-ended output device, which can sense small changes in a magnetic field. The 3501T includes a Hall cell, linear amplifier, emitter-follower output, and a regulator.

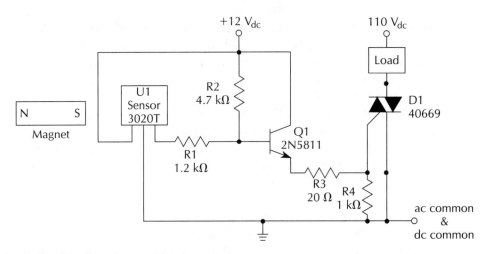

**Fig. 1-25** Hall-effect ac switch.

A ferrous metal detector, shown in Fig. 1-26, can detect metal objects passing the sensor. A magnet is placed with its north pole facing the 3501T sensor. As a metal object passes the detector assembly, the signal is amplified by an LM741C op amp. The output drives a relay. This detector circuit can be used for sensing or counting metal parts, as well as pressure sensing. The output is linear and can provide a continuous display output. The 3501T sensor can also detect the absence of metal objects, as shown in Fig. 1-27. The 3501T can also be used for a tilt-angle sensor, as well as a vibration sensor.

The next Hall-effect sensor is the UGN3501M, which is ideal for accurately measuring and controlling current, velocity, position, weight, and thickness. The UGN3501M is a linear differential-type sensor that includes a Hall cell, a differential amplifier, an emitter follower, and a regulator. Figure 1-28 illustrates a cur-

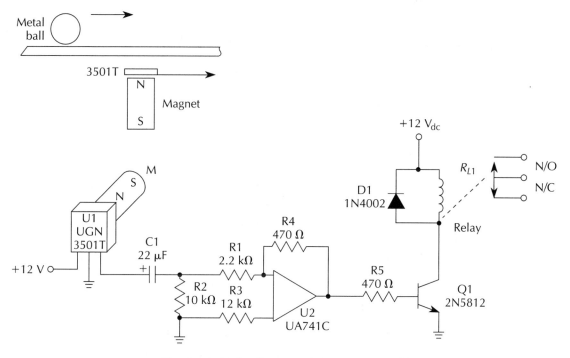

**Fig. 1-26** Hall-effect metal-presence switch.

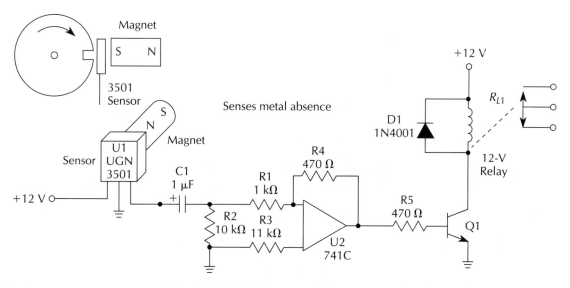

**Fig. 1-27** Hall-effect metal-absence switch.

22  Sensors and Detection Circuits

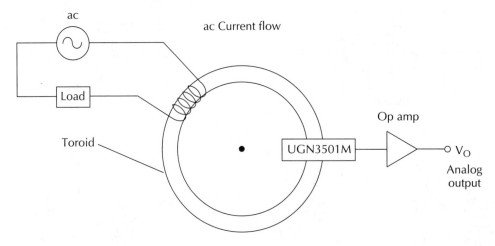

**Fig. 1-28**  *Current-sensing Hall-effect sensor.*

rent-measurement circuit using the UGN3501M sensor. A wire carrying ac current connected in series with a load is wrapped around a toroid. A UGN3501M sensor is placed on the toroid and its output drives a 741 op amp. The resultant output can indicate the varying current when connected to a current meter, or the signal could be applied to an analog-to-digital (A/D) converter card in a personal computer. The differential sensor can also be used as a basic gaussmeter, as shown in Fig. 1-29. As the magnetic field is moved closer to the Hall sensor, an output can be displayed on a voltmeter. A 200-Ω potentiometer zeros the cir-

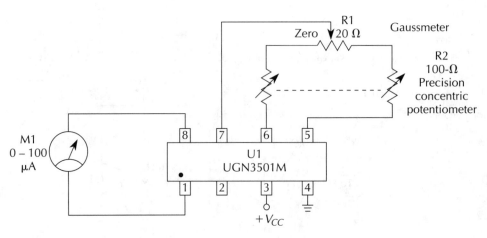

**Fig. 1-29**  *Gaussmeter.*

cuit. This simple gaussmeter is suited to many ac field applications and can be constructed quite inexpensively.

The output of the gaussmeter is on pin 3 and pin 1 of the UGN3501M. The output could also drive an op amp such as an LM741, which will amplify the output of the sensor. A pressure sensor could be fabricated by attaching a linear Hall-effect sensor to a magnet, as shown in Fig. 1-30. As the metal disk travels toward the sensor, a proportional pressure output is displayed. A pressure sensor of this type would be used in a nonmetallic tank or vessel. The linear Hall-effect sensor can also be used for an analog tilt indicator or vibration sensor.

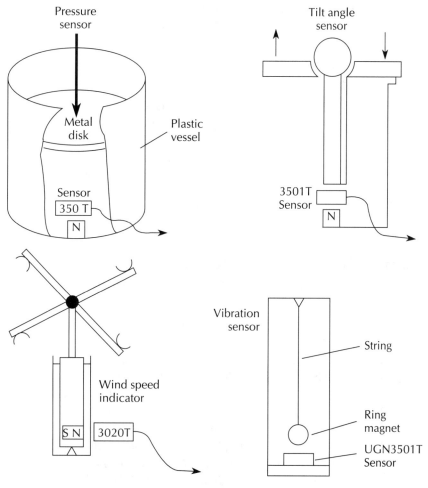

**Fig. 1-30** *Hall-effect sensor applications.*

The Hall-effect sensor can be used in many applications in self-contained systems, or the sensor outputs can be connected to a central alarm panel, such as in a security system.

## Hall-effect switch parts list

| Quantity | Part | Description |
|---|---|---|
| 1 | R1 | 1-kΩ, ¼-W resistor |
| 1 | R2 | 56-Ω, ¼-W resistor |
| 1 | D1 | 1N4001 silicon diode |
| 1 | Q1 | 2N5812 transistor |
| 1 | Q2 | 2N3055 transistor |
| 1 | U1 | UGN3120 Hall cell (Allegro Microsystems) |
| 1 | RY-1 | 9–12-V relay |
| 1 | MAGNET | Bar or cylinder magnet |

## Hall-effect ac switch parts list

| Quantity | Part | Description |
|---|---|---|
| 1 | R1 | 1.2-kΩ, ¼-W resistor |
| 1 | R2 | 4.7-kΩ, ¼-W resistor |
| 1 | R3 | 120-Ω, ¼-W resistor |
| 1 | R4 | 1-kΩ, ¼-W resistor |
| 1 | Q1 | 2N5811 transistor or equivalent |
| 1 | D1 | 40669 triac |
| 1 | U1 | UGN3120 Hall switch (Allegro Microsystems) |

## Hall-effect metal presence switch parts list

| Quantity | Part | Description |
|---|---|---|
| 1 | R1 | 2.2-kΩ, ¼-W resistor |
| 1 | R2 | 10-kΩ, ¼-W resistor |
| 1 | R3 | 12-kΩ, ¼-W resistor |
| 1 | R4 | 470-kΩ, ¼-W resistor |
| 1 | R5 | 470-Ω, ¼-W resistor |
| 1 | C1 | 22-µF, 25-V electrolytic capacitor |
| 1 | D1 | 1N4002 silicon diode |
| 1 | Q1 | 2N5812 transistor |
| 1 | U1 | UGN3503 5-V Hall cell (Allegro Microsystems) |
| 1 | U1 | UGN3501 12-V Hall cell (Allegro Microsystems) |

| 1 | U2 | UA741 op amp |
| 1 | RY-1 | 5–12-V relay |
| 1 | MAGNET | Bar or cyclinder magnet |

**Hall-effect metal absence switch parts list**

| Quantity | Part | Description |
|---|---|---|
| 1 | R1 | 1-k$\Omega$, ¼-W resistor |
| 1 | R2 | 10-k$\Omega$, ¼-W resistor |
| 1 | R3 | 11-k$\Omega$, ¼-W resistor |
| 1 | R4 | 470-k$\Omega$, ¼-W resistor |
| 1 | R5 | 470-$\Omega$, ¼-W resistor |
| 1 | C1 | 1-$\mu$F, 25-V capacitor |
| 1 | Q1 | 2N5812 transistor or equivalent |
| 1 | U1 | UGN3501 12-V Hall cell (Allegro Microsystems) |
| 1 | U1 | UGN3503 5-V Hall cell (Allegro Microsystems) |
| 1 | U2 | UA741C op amp |
| 1 | D1 | 1N4001 silicon diode |
| 1 | RY-1 | 5–12-V SPST relay |
| 1 | MAGNET | Bar or cylinder magnet |

**Gaussmeter parts list**

| Quantity | Part | Description |
|---|---|---|
| 1 | R1 | 20-$\Omega$, ¼-W resistor |
| 1 | R2 | 100-$\Omega$, precision concentric potentiometer |
| 1 | U1 | UGN350M Hall cell (Allegro Microsystems) |
| 1 | M | 0–100-$\mu$A panel meter |

# ac-magnetic field detector

A magnetic field detector is a sensitive indicator of a moving or dynamic ac field. This detector might be of interest if you are concerned about radiation from magnetic fields for safety reasons. The magnetic field detector is shown in Fig. 1-31. The detector can also indicate if a relay is energized or if a high-voltage transformer is operating. Any device with a transformer may be detected readily. The ac-magnetic field detector measures a single-axis field. Therefore, position the detector in each desired axis or build two similar coils and either switch them into the circuit or build three display units to monitor all three axes at once.

## 26 Sensors and Detection Circuits

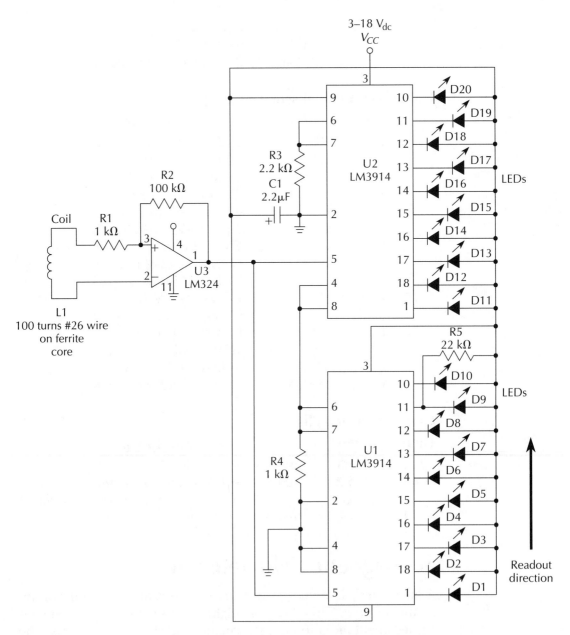

**Fig. 1-31** *Bar-graph ac magnetic field detector.*

The experimental field detector begins with a sense coil. The sense coil consists of 100–200 turns of 28-gauge enameled wire wound on a plastic form covered with aluminum foil and placed inside a plastic film container. The coil is coupled to an LM324 op amp via a 1-kΩ resistor. A 100-kΩ feedback resistor is connected between pins 1 and 3. By switching in different feedback resistors, various sensitivity ranges can be selected. The heart of the ac field meter is two LM3914 DOT/BAR display driver chips, wired so that the first chip drives the second one for a 20-LED sequential display. Two 10-position LED display packages can simplify construction. The LM324 was used because it is a single-supply device. Power is applied to pins 4 and ground to pin 11. The gaussmeter can be powered by a 9-V transistor battery.

Calibration of the ac field meter is accomplished by connecting an ac milliammeter, an 8.2 $V_{ac}$ transformer, and a 100-Ω potentiometer in series with a calibration coil. A calibration coil consists of 110 turns of 26-gauge enameled wire. The sensor coil is placed inside the calibration coil. The calibration formula is: gauss equals turns per meter times the ac amperes of coil current. Then, coil amperes equals gauss divided by the number of turns per meter, multiplied by 20. For example, 110 turns × 20 = 2,200 turns per meter. (Note: a 5-cm coil is 5% of a meter.) Now, 2,200 turns per meter times the ac current in milliamperes equals the field in gauss. The gaussmeter can detect the presence of magnets, but, since it is not a moving field, you have to sweep the magnet or the coil past each other to see an indication.

### Bar-graph ac-magnetic-field meter parts list

| Quantity | Part | Description |
|---|---|---|
| 2 | R1, R4 | 1-kΩ, ¼-W resistor |
| 1 | R2 | 100-kΩ, ¼-W resistor |
| 1 | R3 | 2.2-kΩ, ¼-W resistor |
| 1 | R5 | 22-kΩ, ¼-W resistor |
| 1 | C1 | 2.2-µF, 25-V electrolytic capacitor |
| 1 | L1 | 100 turns 26-gauge enameled wire on ferrite core |
| 2 | U1, U2 | LM3914 display drive ICs |
| 20 | D1-D20 | Red LEDs |

## Bridge sensors

A bridge circuit measures the electrical property of a circuit element indirectly by comparing against a known similar element. The two primary ways of operating a bridge circuit are as a null detector and as a device that directly reads a voltage or current.

Null detectors are primarily used in feedback systems involving electromechanical movement. These systems seek to force the active resistive element, such as a thermistor or strain gauge which is connected to a mechanically coupled potentiometer, to balance the bridge. The null is independent of the excitation voltage.

A basic bridge is shown in Fig. 1-32, and illustrates a null condition when R1/R4 = R2/R3. Figure 1-33 shows a bridge with all resistors equal, but one of them, R1, is a variable active sensing element, such as thermistor or light-sensitive resistor.

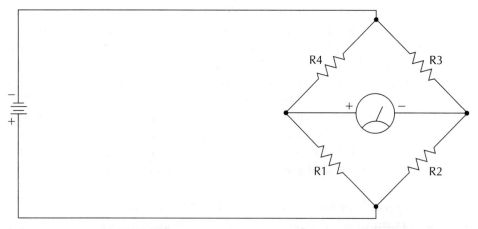

**Fig. 1-32** Simple dc bridge circuit.

The sensitivity of a bridge is the ratio of the excitation voltage to the maximum charge in the output of the circuit. The sensitivity can be doubled if two identical variable elements are used in positions R1 and R3, as shown in Fig. 1-34. An example would be two identical strain-gauge resistive-type sensors aligned in a single pattern. The resultant output would be double.

A special doubling configuration is shown in Fig. 1-35, which consists of four variable resistive sensors elements. Two of the elements increase, i.e., R1 and R3, as the other two elements, R2 and R4, decrease in the same ratio. Two identical two-element strain gauges or sensors are attached to opposite faces of a thin carrier or substrate to measure its bending properties. The output of such a bridge arrangement would be four times the output of a single-element sensor. The complementary resistance changes would result in a linear output.

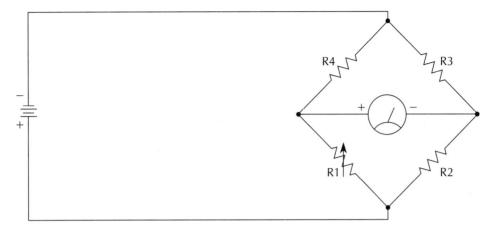

**Fig. 1-33** *Single variable dc bridge.*

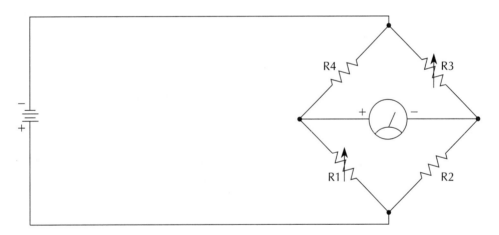

**Fig. 1-34** *Dual variable dc bridge.*

Most instrumentation systems rely on a form of the basic bridge circuit to monitor all types of physical phenomena, using light-sensitive resistors, RTD sensors, and pressure, strain, and flow sensors. Many precision instrumentation systems utilize a precision op amp or chopper-stabilized amplifier. The output of the op amp can be directed to a digital panel meter or an A/D converter card inside a personal computer.

A precision instrumentation bridge circuit is shown in Fig. 1-36. A resistive sensor, R4, unbalances the bridge, producing a

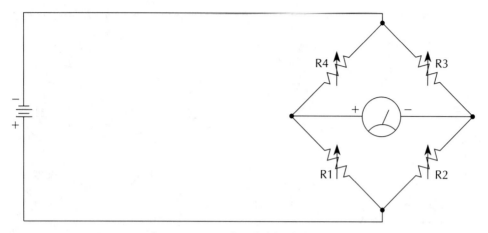

**Fig. 1-35** Quad variable dc bridge.

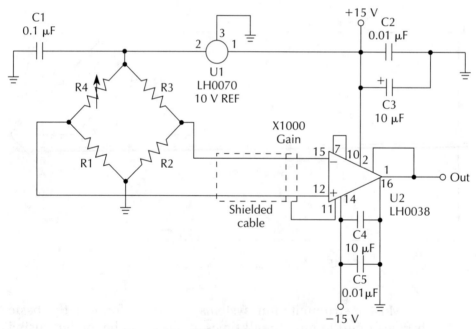

**Fig. 1-36** Instrumentation-grade bridge circuit.

tiny output that is coupled to a National Semiconductor LH0038 three-stage precision instrumentation-amplifier chip with internal gain resistors. The amplifier has a gain factor up to 2000. A 10-V reference produced from the LH0070 is applied to the bridge to provide an accurate source voltage or reference. The system has excellent common-mode rejection and can be used in precision measuring applications. A shielded cable should be used when connecting the sensor if it is not located near the electronics. Any resistive-type sensor could be substituted for a strain gauge, such as a pressure sensor or light-detection circuits. The bridge amplifier is used in many precision measurement circuits, including seismology, water/resource management, ballistics, etc.

### Instrumentation-grade bridge amplifier parts list

| Quantity | Part | Description |
| --- | --- | --- |
| 3 | R1, R2, R3 | 10-k$\Omega$, ¼-W resistor |
| 1 | R4 | 10-k$\Omega$ resistive sensor |
| 2 | C2, C5 | 0.01-$\mu$F, 25-V capacitor (disk) |
| 1 | C1 | 0.1-$\mu$F, 25-V capacitor |
| 2 | C3, C4 | 10-$\mu$F, 25-V electrolytic capacitor |
| 1 | U1 | LH0070 10-V reference (National Semiconductor) |
| 1 | U2 | LH0038 instrumentation amplifier |

## Maxwell bridge

The basic bridge circuit is not limited to resistive dc circuits. A device called the Maxwell bridge can measure an unknown capacitance or inductance. A Maxwell bridge is shown in Fig. 1-37. IC1 and IC2 form an oscillator that can be made to oscillate between 1 and 10-kHz. A small 3:1 ratio transformer couples the drive circuit to the measuring circuit. An unknown coil value can be determined by placing the coil at LX. The value of an unknown capacitor can be measured by connecting the capacitor to A and B.

Generally, an analog meter would be connected to points x and y as shown. The null value can then be determined on the meter. Calibration is performed with RA and RB. A null or fine tuning is then accomplished with potentiometer R3.

We can construct a metal detector or automobile/truck sensor with this bridge. A coil or pickup device can be constructed by winding 75 to 100 turns of 26-gauge enameled wire, random wound on a 1-foot square coil form. The coil should be constructed so that it is protected from the elements, because it is intended to be buried in a driveway. Connect an LM311 comparator, as shown, to points x and y. The LM311 is coupled to a 2N3904 npn transistor, which can drive a small relay (Radio Shack 275-240). After connecting the coil, a null first must be obtained by ad-

## 32 Sensors and Detection Circuits

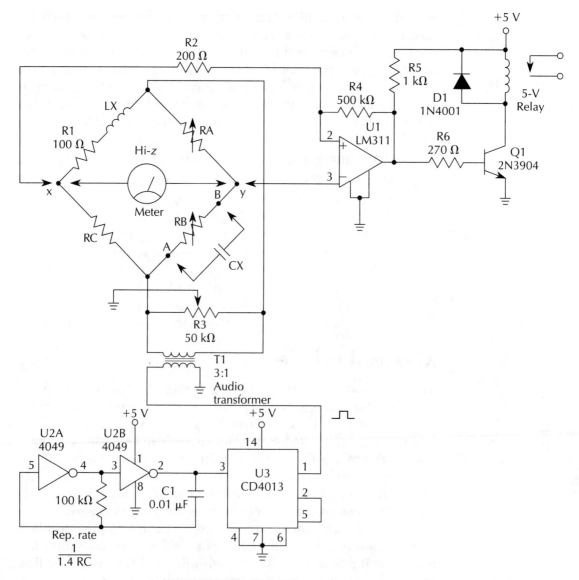

**Fig. 1-37** Maxwell ac bridge circuit.

justing RA and RB. A final null is made with R3 as described previously. R3 is adjusted so no output is present at points x and y. As a large metal object or car approaches, the null condition will become upset and the resultant change is then applied to the comparator, which could activate a buzzer. The metal detector could also be used to locate hidden pipes in the ground, by carrying the coil on a plastic pole and holding the coil parallel to the ground as you search the suspected area.

## Maxwell ac/dc bridge parts list

| Quantity | Part | Description |
|---|---|---|
| 1 | RA | 100-$\Omega$, ½-W potentiometer |
| 1 | RB | 100-$\Omega$, ½-W potentiometer |
| 1 | RC | 100-$\Omega$, ½-W resistor |
| 1 | R1 | 100-$\Omega$, ½-W resistor |
| 1 | R2 | 200-$\Omega$, ½-W resistor |
| 1 | R3 | 50-k$\Omega$ potentiometer (trim) |
| 1 | R4 | 500-k$\Omega$, ¼-W resistor |
| 1 | R5 | 1-k$\Omega$, ¼-W resistor |
| 1 | R6 | 270-$\Omega$, ¼-W resistor |
| 1 | R7 | 100-k$\Omega$, ¼-W resistor |
| 1 | LX | Known if unknown CX |
| 1 | CX | Known if unknown LX |
| 1 | C1 | 0.01-$\mu$F, 25-V capacitor (disk) |
| 1 | D1 | 1N4001 silicon diode |
| 1 | Q1 | 2N3904 transistor |
| 1 | U1 | LM311 op amp (National Semiconductor) |
| 1 | U2 | CD4049 CMOS HEX inverter |
| 1 | U3 | CD4013 CMOS flip-flop |
| 1 | T1 | 3:1 miniature transformer |
| 1 | M | 20-k$\Omega$/V multimeter or panel meter |
| 1 | RY-1 | 5-V relay |

## Pyroelectric detector

The pyroelectric or body-heat, detector can sense humans or large animals up to 50 feet away. The pyroelectric or passive infrared sensor requires no external field excitation, as needed by ultrasonic or microwave detectors. Only a human IR source is needed to trigger a pyroelectric sensor. Pyroelectric detectors are inexpensive and very reliable and have become the preferred alarm sensor in the security industry. When used with a discriminator circuit, the systems become virtually free of false alarms. The coverage of the pyroelectric sensor can be either short range, i.e., an 8 to 10-foot-square area in front of the sensor, or long range out to 50 feet but with a narrow beamwidth. A small plastic Fresnel lens placed in front of the sensor determines the actual range of the sensor (see Fig. 1-38). The sensor can be used without a lens as a touch switch to detect human hands or fingers.

A number of plastics and unsymmetrical crystals, such as lithium tantalate, can provide the pyroelectric effect. A thin wafer

34   Sensors and Detection Circuits

**Fig. 1-38**  *Pyroelectric sensor and lens.*

of this material is sandwiched between two electrodes, and an internal field is generated and the charge across the crystal can be monitored. To detect the charge, a high-impedance field-effect transistor and an internal matching resistor are placed together with the lithium-tantalate crystal in a TO-5 size transistor case. A sense window in front of the crystal is covered with an infrared filter. The internal FET amplifies the incoming signal, which is fed directly to a 741 op amp, as shown in Fig.1-39. The ELTEC 406 series single-crystal sensor is coupled to the inverting input of the op amp on pin 2 (see Fig. 1-40). When the voltage level on pin 2 has exceeded the threshold voltage on pin 3, which is controlled by a 100-k$\Omega$ potentiometer, an output signal on pin 6 drives a small relay via a 2N2222 npn transistor. The output can directly drive a buzzer or can be used to trigger a latching alarm by substituting the transistor with an SCR. The circuit can also activate a central alarm panel. Other sensor configurations are available, including dual sensors configured in parallel or opposed parallel for increased discrimination and stability. Flame sensors are also available, which can detect pilot lamps in heaters or furnaces. Samples of the devices are available upon request by giv-

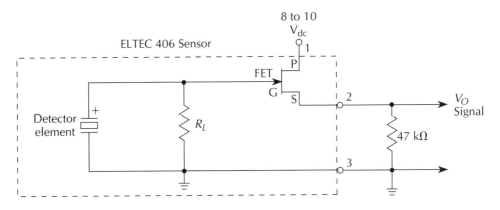

**Fig. 1-39** *Pyroelectric sensor diagram.*

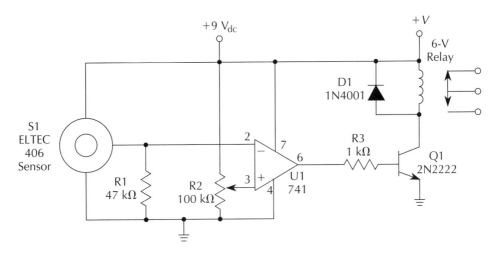

**Fig. 1-40** *Pyroelectric sensor and switch circuit.*

ing your company name and application. A pyroelectric detector with dual opposed parallel detectors for increased discrimination is described with a construction project in the last chapter.

### Pyroelectric sensor and switch circuit parts list

| Quantity | Part | Description |
|---|---|---|
| 1 | R1 | 47-kΩ, ¼-W resistor |
| 1 | R2 | 100-kΩ, ¼-W resistor |

| Quantity | Part | Description |
|---|---|---|
| 1 | R3 | 1-kΩ, ¼-W resistor |
| 1 | D1 | 1N4001 silicon diode |
| 1 | Q1 | 2N2222 pnp transistor |
| 1 | U1 | UA741C op amp |
| 1 | RY-1 | 6-V SPST relay |
| 1 | S1 | ELTEC 406 pyroelectric sensor |

# Pressure sensor

Pressure sensors can be used to measure a variety of physical parameters, including wind speed, liquid velocity, barometric pressure, altitude, and the speed of rotational objects such as bicycles and automobiles. Pressure sensing in the past has required very costly sensors, but recently many new sensitive, low-cost models have become available. A low-cost pressure sensor is shown in Fig. 1-41. A new catalog from Sensym offers many different types of pressure sensors that can measure absolute, differential, and gauge pressures.

*Fig. 1-41*  Pressure sensor.

Absolute-pressure sensors measure changes in barometric pressure and are commonly used in altimeters. These applications require a reference to a fixed pressure and cannot be simply referenced to the ambient pressure. Absolute pressure is defined as the pressure measured relative to a perfect vacuum; i.e., 10 psi is 10 pounds per square inch above a perfect vacuum.

Differential pressure is the pressure difference measured between two pressure sources. When one source is used for measurement, the ambient source is known as relative or gauge pressure, measured in pounds per square inch or psig. Gauge pressure is simply a special case of differential pressure, with pressure measured differently but always relative to the local ambient pressure. A simple sensitive pressure switch is shown in Fig. 1-42.

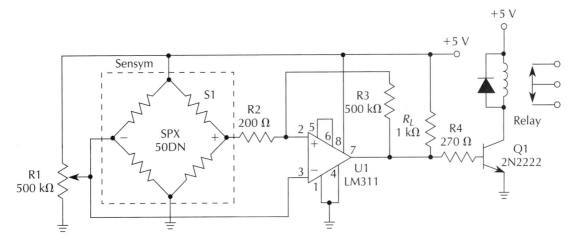

**Fig. 1-42** *Pressure switch.*

The pressure switch uses a Sensym SPX500N sensor bridge coupled to an LM311 comparator via a 200-Ω resistor. A 500-kΩ potentiometer is the sensitivity or threshold control. The output of the comparator on pin 7 is directed to Q1, a 2N2222 transistor. The output transistor can drive a low-current relay, which can activate a motor or buzzer or trigger an alarm panel.

Many pressure applications require an analog output or display. Figure 1-43 illustrates an analog pressure system that drives an analog LED bar-graph display. An analog or digital meter module could be substituted for the bar graph if desired. Many low-cost digital panel meters or modules are available from suppliers listed in the Appendix. A chart is provided to help select the proper sensor and feedback resistors for your application. Two resistors, Rg and Rp, select the feedback or gain parameters.

An instrument-grade pressure measurement system is shown in Fig. 1-44. This system uses a precision-grade bridge sensor with a precise voltage reference source coupled to the bridge. The signal is amplified and processed by A1-A4, four Linear Technology LT1014 integrated op amps. The analog voltage output of A3 is very accurate. The precision output is connected directly to an ADC0804 8-bit analog-to-digital (A/D) converter, which can be

## 38 Sensors and Detection Circuits

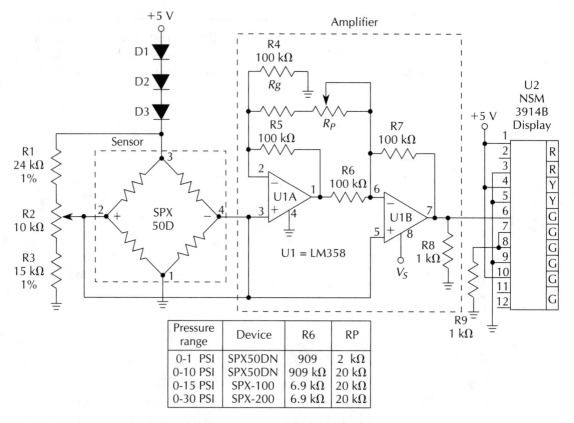

**Fig. 1-43** *Analog bar-graph pressure sensor.*

connected to an 8-bit parallel input on a personal computer. The analog output at A3 could also be applied to an A/D converter card placed inside a computer, to act as a multichannel pressure data logger by combining a number of sensor channels.

## Pressure switch parts list

| Quantity | Part | Description |
|---|---|---|
| 1 | R1 | 500-k$\Omega$ potentiometer |
| 1 | R2 | 200-$\Omega$, ¼-W resistor |
| 1 | R3 | 500-k$\Omega$, ¼-W resistor |
| 1 | R4 | 270-$\Omega$, ¼-W resistor |
| 1 | RL | 200-$\Omega$, 1-k$\Omega$, ½-W resistor |
| 1 | D1 | 1N4002 silicon diode |
| 1 | U1 | LM311 op amp |
| 1 | Q1 | 2N2222 pnp transistor |
| 1 | RY-1 | 5-V SPST relay |
| 1 | S1 | Sensym SPX50DN pressure sensor |

Pressure sensor 39

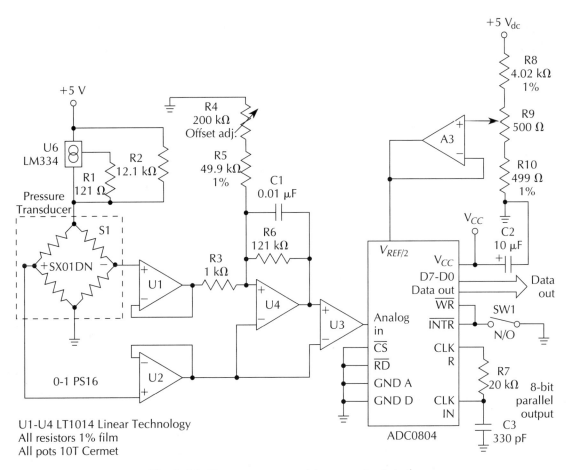

**Fig. 1-44** *Pressure sensor with computer interface.*

### Analog bar-graph pressure sensor parts list

| Quantity | Part | Description |
|---|---|---|
| 1 | R1 | 24-kΩ, 1%, ¼-W resistor |
| 1 | R2 | 10-kΩ, ¼-W potentiometer |
| 1 | R3 | 15-kΩ, 1%, ¼-W resistor |
| 1 | R4, R5, R6, R7 | 100-kΩ, ¼-W resistor |
| 1 | R8 | 1-kΩ, ¼-W resistor |
| 1 | R9 | 3.3-kΩ, ¼-W resistor |
| 1 | Rp | See table (scaling) |
| 1 | Rg | See table (scaling) |
| 1 | U1 | LM358 op amp |
| 1 | U2 | NSM3914B LED dot display driver |
| 1 | S1 | Sensym SPX50DN pressure sensor |

## Pressure sensor with computer interface parts list

| Quantity | Part | Description |
|---|---|---|
| 1 | R1 | 121-Ω, 1%, ¼-W resistor |
| 1 | R2 | 12.1-kΩ, 1%, ¼-W resistor |
| 1 | R3 | 1-kΩ, ¼-W resistor |
| 1 | R4 | 200-kΩ, ¼-W resistor |
| 1 | R5 | 49.9-kΩ, 1%, ¼-W resistor |
| 1 | R6 | 121-kΩ, ¼-W resistor |
| 1 | R7 | 20-kΩ, ¼-W resistor |
| 1 | R8 | 4.02-kΩ, 1%, ¼-W resistor |
| 1 | R9 | 500-Ω trim pot |
| 1 | R10 | 499-Ω, 1%, ¼-W resistor |
| 1 | C1 | 0.01-μF, 25-V capacitor (disk) |
| 1 | C2 | 10-μF, 25-V electrolytic capacitor |
| 1 | C3 | 330-pF, 25-V capacitor (disk) |
| 1 | U1-U5 | LT101A op amp (Linear Technology) |
| 1 | U7 | ADC0804 A/D IC (National Semiconductor) |
| 1 | U6 | LM334 op amp (National Semiconductor) |
| 1 | SW-1 | SPST toggle switch |
| 1 | S1 | Sensym SX01DN pressure sensor |

# Toxic-gas sensor

A danger always exists when combustible gases such as propane or gasoline are confined in a small area. The toxic-gas alarm shown in Fig. 1-45 uses a tin-oxide-semiconductor sensor. A thin filament coil is heated from a 12-V battery via IC1 and IC2, which pulses the voltage to the sensor coil, thus saving a considerable amount of energy. Zener diode D1 provides a constant filament-supply voltage to the sensor coil. The sensor's resistance lowers as the sensor is exposed to toxic gases, such as hydrogen, carbon monoxide, and propane. As the sensor's resistance decreases, the gate voltage to the SCR increases. When the threshold voltage on the gate is reached, the SCR fires and an alarm buzzer is activated. Once triggered, the buzzer sounds and switch S1 must be used to reset the alarm. Because the sensor has a fair amount of thermal inertia, S1 should be in the OFF or OPEN position for about three or four minutes upon initial power-up, allowing the sensor to stabilize, thus preventing false alarms. The sensitivity control R7 is adjusted to the desired value prior to triggering the SCR.

The toxic-gas alarm is sensitive to less than 100 ppm of carbon monoxide. This simple gas alarm is useful for boats, basements,

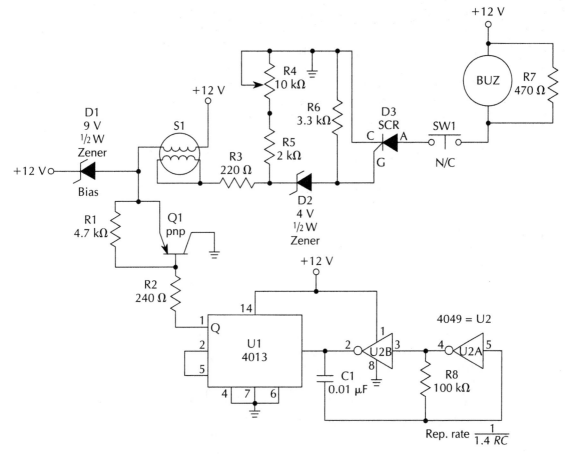

**Fig. 1-45** *Toxic-gas sensor.*

and cabins. It could save a life. Some of the companies listed in the Appendix offer plans and kits for various toxic-gas sensors.

## Toxic-gas sensor parts list

| Quantity | Part | Description |
|---|---|---|
| 1 | R1 | 4.7-k$\Omega$, ¼-W resistor |
| 1 | R2 | 240-$\Omega$, ¼-W resistor |
| 1 | R3 | 220-$\Omega$, ¼-W resistor |
| 1 | R4 | 10-k$\Omega$, ¼-W potentiometer |
| 1 | R5 | 2-k$\Omega$, ¼-W resistor |
| 1 | R6 | 3.3-k$\Omega$, ¼-W resistor |
| 1 | R7 | 470-$\Omega$, ¼-W resistor |
| 1 | R8 | 100-k$\Omega$, ¼-W resistor |

| Quantity | Part | Description |
|---|---|---|
| 1 | C1 | 0.01-µF, 25-V capacitor (disk) |
| 1 | D1 | 9-V zener diode (NTE139A) |
| 1 | D2 | 4-V zener diode (NTE5068A) |
| 1 | D3 | SCR (NTE5408) |
| 1 | Q1 | 2N2222 transistor |
| 1 | U1 | CD4013 CMOS flip-flop |
| 1 | U2 | CD4049 CMOS hex inverter |
| 1 | BUZ | 6–9V piezo buzzer |
| 1 | S1 | Toxic-gas sensor TGS203 (Figaro or equivalent) |

# Humidity sensor

The recently introduced *hygristor* allows the design of low-cost, rapid-response, humidity-sensing devices. A hygristor is a sensing device that changes its electrical resistance with changes in the ambient humidity. Typical applications for the hygristor include portable weather instruments, weather sondes, greenhouse sensors, environmental sensors, dew-point sensors, and soil and agriculture monitoring devices. Our sample hygristor, shown in Fig. 1-46, was a 2-inch × ½-inch deposited-film, plastic-type sensor that measures relative humidity from 0 to 100% within a 2-second response time. The hygristor operates from −40°C to +50°C, with an operational resistance that varies from 10 kΩ to 100 kΩ.

**Fig. 1-46** *Humidity sensor.*

The circuit shown in Fig. 1-47 depicts a typical hygristor connected to an ohms-to-voltage converter. The humidity sensor is coupled through a l00-kΩ resistor to one half of an LF353 op amp. Pin 3 of the op amp is clamped via the two 1N914 diodes. The output of section A of the op amp is then fed to section B via pin 5. A

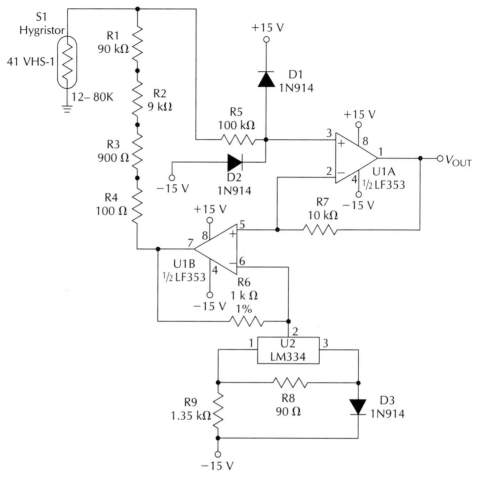

**Fig. 1-47** Humidity sensor diagram.

100-k$\Omega$ resistive ladder couples the sensor to pin 7 of the LF353. A compensation network is connected to pin 6 of the IC. Note the 1-k$\Omega$, 1% resistor between pins 6 and 7. The output of the op amp on pin 1 is a dc voltage that can be read with a digital panel meter, a data logger, or an A/D card placed in a computer. A plus-and-minus power supply is required for this circuit.

The manufacturer of the hygristor offers various custom resistance values and a variety of mechanical packages. Calibrating the humidity sensor circuit and converter can be accomplished by using a known reference source. Samples of the hygristors are available directly from the manufacturer listed in the Appendix.

## Humidity sensor parts lists

| Quantity | Part | Description |
|---|---|---|
| 1 | R1 | 90-k$\Omega$, 1%, ¼-W resistor |
| 1 | R2 | 9-k$\Omega$, 1%, ¼-W resistor |
| 1 | R3 | 900-$\Omega$, 1%, ¼-W resistor |
| 1 | R4 | 100-$\Omega$, 1%, ¼-W resistor |
| 1 | R5 | 100-k$\Omega$, 1%, ¼-W resistor |
| 1 | R6 | 1-k$\Omega$, 1%, ¼-W resistor |
| 1 | R7 | 10-k$\Omega$, 1%, ¼-W resistor |
| 1 | R8 | 90-$\Omega$, 1%, ¼-W resistor |
| 1 | R9 | 1.35-k$\Omega$, 1%, ¼-W resistor |
| 3 | D1, D2, D3 | 1N914 silicon diode |
| 1 | U1 | LF353 op amp (National Semiconductor) |
| 1 | U2 | LM334 op amp (National Semiconductor) |
| 1 | S1 | 41-VHS-1 hygristor (Victory Engineering) |

# Fiberoptic sensing

Counting and position sensing can be implemented using a bifurcated fiberoptic light pipe (see Figs. 1-48 and 1-49). Two polished fiberoptic pipes are fused together at the junction of the two light pipes. A single emerging light pipe is used as a sensing head, which can be used to count very small parts or calibrated markings or rulings such as microscope gradiations. The fiberoptic sensor can be used to keep track of both linear and rotary tables or stages.

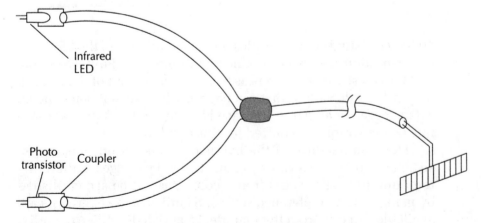

***Fig. 1-48*** *Bifurcated light-pipe drawing.*

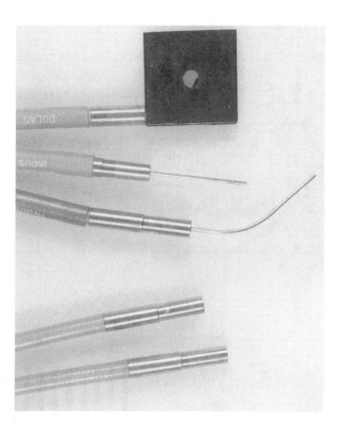

**Fig. 1-49**  Bifurcated light pipes.

An infrared LED is placed at one end of the two light pipes as shown. A phototransistor is then placed at the other end of the two light pipes. The output of the phototransistor is coupled to a light-detector interface, shown in Fig. 1-50. The phototransistor is fed to the 741 op amp, which amplifies the phototransistor's output signal. The output of the op amp is then passed to a 555 timer IC, which outputs a precise 5-V pulse. The output of this pulse shaper can be sent to an I/O counter card placed in a personal computer.

To provide direction or position data, a second light-pipe assembly would be needed. The two light-head assemblies are placed next to each other. One sensing head would be aligned on an opaque ruled line. The second light-pipe head assembly would be mounted precisely in between the next ruling or marking, as shown in Fig. 1-51. By using the two light-pipe assemblies, we have created a rotary optical encoder with quadrature output, which can sense direction and counts in one operation. The two light-interface outputs could be connected to the input of the rotary optical interface to obtain steering logic, which determines directional data.

46  Sensors and Detection Circuits

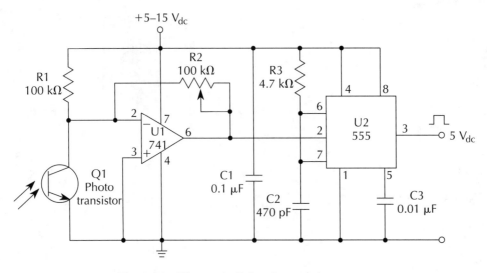

**Fig. 1-50** *Fiberoptic light-pipe receiver.*

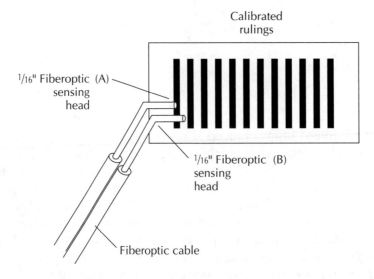

**Fig. 1-51** *Direction-sensing light pipes.*

As noted earlier, if you were to obtain a more expensive type of PC counting card, this steering-logic circuit could be eliminated and the light interfaces could be coupled directly to the counter card, which has built-in quadrature-input sensing. The bifurcated light pipes are available in many configurations from ⅛-inch round and square types to the long-nosed micro ¹⁄₁₆-inch-type heads. These fiberoptic sensing heads are ideal for micropositioning, optical microscope stages, etc.

### Fiberoptic light-pipe receiver parts list

| Quantity | Part | Description |
|---|---|---|
| 1 | R1 | 100-kΩ, ¼-W resistor |
| 1 | R2 | 100-kΩ, ¼-W potentiometer |
| 1 | R3 | 4.7-kΩ, ¼-W resistor |
| 1 | C1 | 0.1-µF, 25-V capacitor |
| 1 | C2 | 470-pF, 25-V capacitor (disk) |
| 1 | Q1 | Phototransistor (NTE3031) |
| 1 | U1 | UA741 op amp |
| 1 | U2 | 555 IC timer |

## Optical rotary encoder

The optical rotary encoder is ideal for digital panel controls or in position-sensing applications where long life, reliability, high resolution, and precise linearity are critical. Rotary optical encoders are often used in computer-aided design/computer-aided manufacturing (CAD/CAM) counting circuits, instrumentation, motor speed controls, and machine tools. The Bourns EN series optical encoder, as shown in Figs. 1-52 and 1-53, is a 1-inch-square device that converts rotary input into electrical signals that can be used by microprocessors without the use of A/D converters.

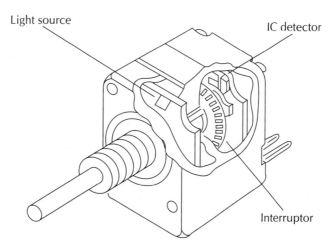

*Fig. 1-52* Optical rotary encoder.

Optical rotary encoders (OREs) combine an LED light source, a photo interrupter disk, and an IC light receiver. Light passing through the slots in the interrupter are sensed and translated into

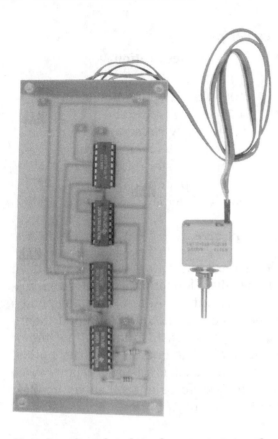

**Fig. 1-53** Encoders and interface.

digital pulses by the phototransistor detector. The ORE has no contacts or wipers that create electrical noise, and they can be operated at high speeds. The EN series encoders use a standard ¼-inch shaft for easy panel mounting.

The ORE provides a quadrature, or two-channel output, as shown in Fig. 1-54. Channel A is shown leading channel B by 90 electrical degrees. ORE outputs are compatible with both CMOS and TTL systems. Up to 256 quadrature output cycles per shift revolution are provided with this system. The quadrature output allows the ORE to determine direction and produce counting pulses with additional electronics, as shown in Fig. 1-55. The quadrature output of the ORE's are 5 $V_{dc}$. The A and B signal outputs from the encoders are coupled to two Schmitt triggers that, in turn, feed three NAND gates. Gates A and B connect to a 7474 flip-flop that provides direction output. NAND gate C of the steering logic connects to a Schmitt gate and then to a series of inverters to provide a small time delay. The output from the 74LS04 is then available to provide counting pulses. The resultant 5-V pulses can drive an I/O card in a personal computer.

## Optical rotary encoder 49

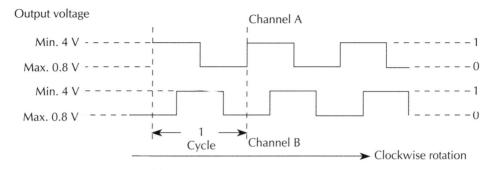

**Fig. 1-54** *Quadrature-output logic diagram.*

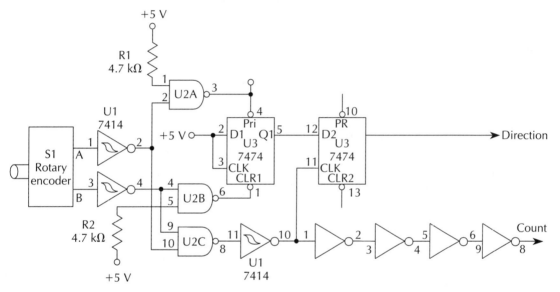

**Fig. 1-55** *Rotary-encoder steering-logic diagram.*

Counting cards are available to direct rotary encoder input, thus eliminating the interface shown. The Bourns optical rotary encoders are available in various resolutions and mounting assemblies. Optical rotary encoders are used extensively in position-sensing applications where a gear could be pieced on the encoder shaft, allowing it to interface with moving tables, rotating shifts, or stepper motors.

### Rotary-encoder steering logic parts list

| Quantity | Part | Description |
|---|---|---|
| 2 | R1, R2 | 4.7-kΩ, ¼-W resistor |
| 1 | U1 | 7414 hex Schmitt trigger |

| Quantity | Part | Description |
|---|---|---|
| 1 | U2 | 7400 quad NAND gate |
| 1 | U3 | 7474 flip-flop |
| 1 | U4 | 7404 hex inverter |
| 1 | S1 | Bourns optical rotary encoder |

## Film strip position sensor

A low-cost film strip position sensor is easy to construct. It can be used to make measurements on devices that will not tolerate loading or drag. Most sensors such as linear variable differential transformers (LVDTs) and resistive encoders inherently cause loading. The film position sensor can be placed in very small spaces, where other types of sensors would be too large, as in micropositioners or microscope stages.

The principle of the film strip position sensor is quite simple. First construct the film strip, which varies linearly in opacity along its length. Then attach the film strip to a moving part and position it so it can modulate the light beam of a slotted optical switch. Generally, it is quite difficult to make a film strip with precise linearly varying density. However, a simple solution to this problem consists of a pattern of opaque bars spaced on 0.01-inch centers, as shown in Fig. 1-56. Varying the width of the spacing from 0 to 100% from one end of the strip to the other essentially creates an optical-pulse-width-modulated output. The 0.01-inch spacing appears to be adequate for this type of detector, because the LED and phototransistor dice or substrate are about 0.05 inch apart.

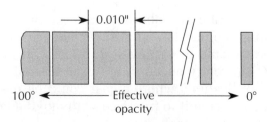

**Fig. 1-56** *Film strip position sensor.*

The film-strip pattern can be manually drawn or computer generated on paper or mylar and then photographically reduced. A high-quality image is not required for this sensor to operate.

Figure 1-57 illustrates placement of the film strip in a typical optical switch assembly. The width of the slot is 0.01 inch; therefore mechanical tolerance is not critical. The electronic interface for the film strip position sensor is shown in Fig. 1-58. An LM317L voltage regulator provides a constant 20 mA of current to the LED in the optical switch. At this level, the full-scale current of the phototransistor may vary from 0.2 mA to 1 mA. The 5-k$\Omega$ potentiometer is adjusted to provide 1 V full scale. To obtain the best linearity for the system, the voltage across the 5-k$\Omega$ potentiometer should be kept to a low value. The voltage from the 5-k$\Omega$ pot is then sent to an LM358 op amp. The voltage is amplified by the op amp to provide a full 0–10 $V_{dc}$ output. The gain of the op amp is set by the ratio of RA and RB. The output of the interface can then be connected to a digital panel meter or an A/D card placed in a personal computer. The film strip position sensor is ideal for miniature devices that can be loaded by external components and where low cost is of great importance.

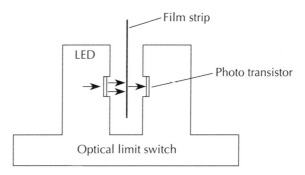

**Fig. 1-57** *Optical position switch.*

## Optical-position light-wave receiver parts list

| Quantity | Part | Description |
|---|---|---|
| 1 | R1 | 62-$\Omega$, ¼-W resistor |
| 1 | R2 | 5-k$\Omega$ potentiometer |
| 1 | RA | 1-k$\Omega$, ¼-W resistor |
| 1 | RB | 9.1-k$\Omega$, ¼-W resistor |
| 1 | U1 | LM317L op amp (National Semiconductor) |
| 1 | U2 | LM358 op amp (National Semiconductor) |
| 1 | OC-1 | M5T-8 optical coupler |

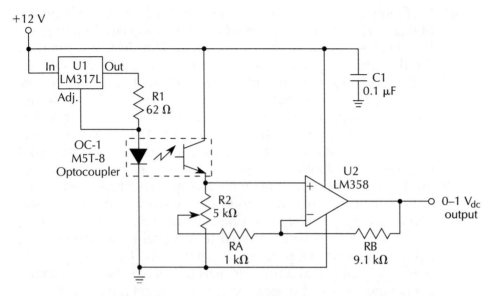

**Fig. 1-58** *Optical position light-wave receiver.*

# Tiltmeter

Electronic tiltmeters and levels have been around for about 40 years. However, most people know little about them and most have seldom seen or used them in sensing applications. Figure 1-59 illustrates a simple electronic level. This novel level uses two common mercury switches, two LEDs, and two penlight cells. Mount the two mercury switches on a 1½- x-6-inch-long,

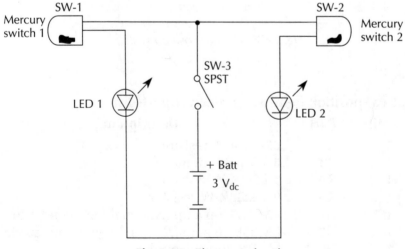

**Fig. 1-59** *Electronic level.*

absolutely flat board, one at each end of the board. The LEDs, switch, and batteries are mounted in a small plastic box attached to the center of the level board. Place the level on an unknown surface. If one LED lights, the surface is not level. If both LEDs light, the surface is level. This type of on-off level is often used to level or position large machinery.

Electrolytic levels and tiltmeters have been around since the late 1930s and early 1940s. They were originally developed for bomb-sighting equipment and aircraft navigation. In recent years, electrolytic or bubble tiltmeters have been used in drilling equipment, seismic sensors, crane leveling, and navigation systems.

The electrolytic cell is composed of a glass cell or enclosed tube with three internal platinum electrodes. The platinum electrodes are partially sheathed with a thin glass bead. The bead insulates a portion of the electrodes to ensure that the bubble will ride on the undistorted portion of the vial, because the vials are usually curved. The glass vials are hand constructed. After the electrodes are installed, the vial is partially filled with a conductive fluid (electrolyte) and then hermetically sealed. A typical electrolytic cell is shown in Fig. 1-60. The glass cell is constructed in such a way that one of the electrodes is always immersed in the electrolyte. This reference electrode is usually placed at the bottom of the vial. The remaining two electrodes are placed on the top of the cell so that either electrode is immersed in liquid, depending on the tilt angle. Thus, a change is established between either electrode and the reference electrode. The many variations of the electrolytic-bubble tiltmeters have slightly different characteristic curves and linearities, including the precision toroid tiltmeter cell, which is highly resistant to vibration and movement and is often used in helicopters for attitude control.

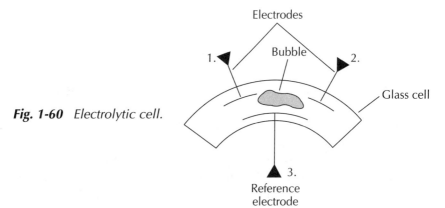

***Fig. 1-60*** *Electrolytic cell.*

One method of measuring tilt angles is to use an LVDT (linear variable differential transformer) signal-conditioner integrated circuit paired with an electrolytic tiltmeter cell (see Figs. 1-61 and 1-62. The heart of the precision tiltmeter is the Analog Devices AD598 integrated circuit. This signal-conditioner chip is commonly used with a coil assembly to measure precise linear displacement. The tiltmeter circuit shown substitutes the electrolytic cell for the coil assembly. The 20-pin LVDT signal-conditioner chip contains a low-distortion sine-wave oscillator, a decoder, divider, summer, filter, and amplifier. The sine-wave output drives the electrolytic cell. The decoder then determines the ratio of difference between the signals on pins 10 and 11. The signals are then divided and summed into one signal that is then filtered and amplified. A precise dc voltage output is proportional to the tilt-angle of the electrolytic cell. The tilt-angle output voltage is presented on pins 16 and 17. This novel circuit can be combined with a digital voltmeter, or the circuit could be connected to a data logger or to an A/D card placed in a personal computer.

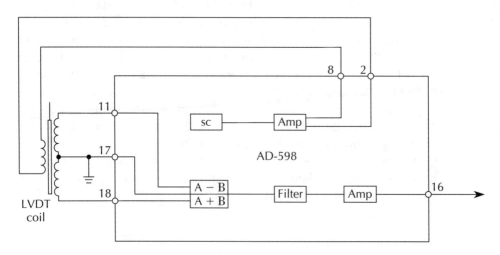

**Fig. 1-61**  *LVDT signal-conditioner chip.*

A simple magnetic tiltmeter can be fabricated using a small magnet suspended from a movable cylinder that is free to swing to and fro. The cylinder is attached to both ends of a small box via a bearing set on each side of the box, as shown in Fig. 1-63. The magnet is allowed to swing over a linear Hall-effect sensor. The Hall-effect sensor, an Allegro Systems UGN3501, consists of a Hall cell, a linear amplifier, a voltage follower, and a voltage regulator, all packaged in a three-lead device. The Hall-effect

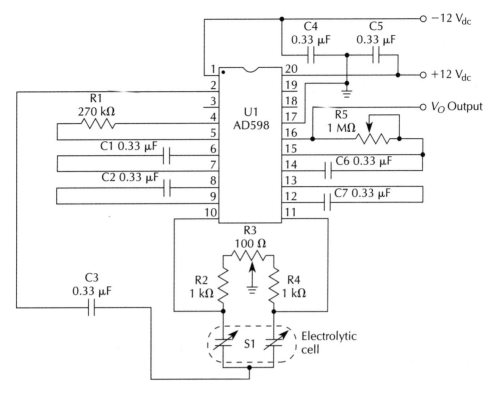

**Fig. 1-62** *LVDT tiltmeter system.*

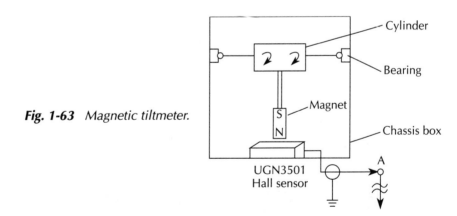

**Fig. 1-63** *Magnetic tiltmeter.*

sensor is coupled to an LM741 op amp as shown in Fig. 1-64. The output of the op amp provides a proportional dc voltage that depends upon the tilt of the magnet over the Hall cell.

Yet another approach to measuring tilt angles uses a micropower LED suspended over a position-sensitive detector (PSD) mounted in a small light-tight enclosure (see Figs. 1-65 and 1-66).

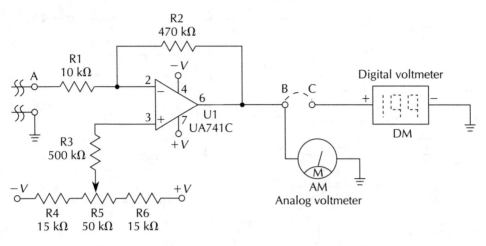

**Fig. 1-64** *Magnetic-tiltmeter electronic circuit diagram.*

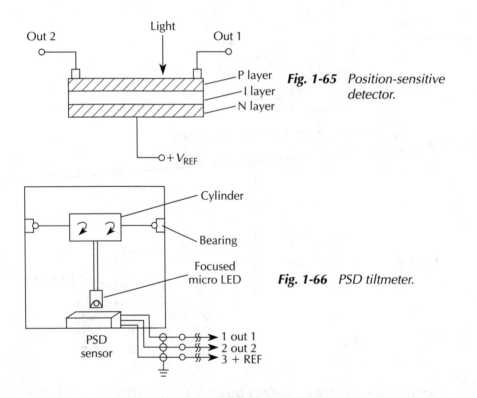

**Fig. 1-65** *Position-sensitive detector.*

**Fig. 1-66** *PSD tiltmeter.*

The PSD is a linear-displacement sensor, which produces a dc voltage proportional to position of the light source over the sensor. This device is discussed elsewhere in this book. The PSD displacement sensor is a three-lead device that requires some additional support circuitry, as shown in Fig. 1-67. The op-amp circuit

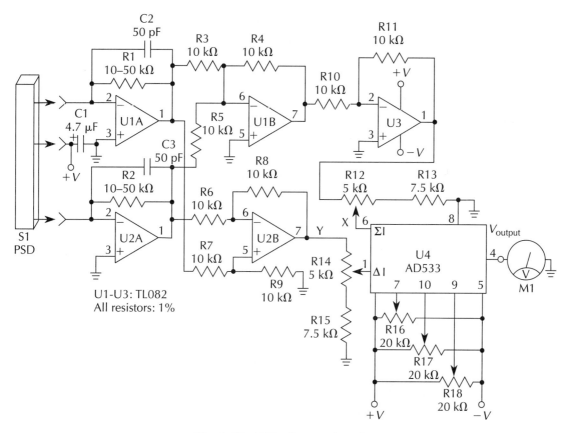

*Fig. 1-67* PSD-tiltmeter interface.

uses an Analog Devices AD533 multiplier/divider integrated circuit. The 10-kΩ to 50-kΩ input resistors are adjusted per your particular circuit parameters. This circuit requires 1% resistors for precision output voltages. The output of the PSD signal conditioner yields a proportional dc voltage between 1 and 10 V. The output could be coupled to a digital meter or A/D card if desired.

Other approaches to measuring tilt angles include resistive displacement and capacitive measurements. Tiltmeters are used in many applications all around us, but they are seldom visible and are therefore taken for granted.

### Electronic level parts list

| Quantity | Part | Description |
|---|---|---|
| 2 | SW-1, SW-2 | Glass mercury switches |
| 1 | SW-3 | SPST toggle switch |
| 2 | LED-1, LED-2 | Red LEDs |

| Quantity | Part | Description |
|---|---|---|
| 2 | BATT | AA penlight cells |
| 1 | BH | Battery holder for two batteries |

## Tiltmeter system parts list

| Quantity | Part | Description |
|---|---|---|
| 1 | R1 | 270-k$\Omega$, ¼-W resistor |
| 2 | R2, R4 | 1-k$\Omega$, ¼-W resistor |
| 1 | R3 | 100-$\Omega$, ¼-W potentiometer |
| 1 | R5 | 1-M$\Omega$, ¼-W potentiometer |
| 7 | C1-C7 | 0.33-µF, 25-V capacitor |
| 1 | U1 | AD598 LVDT signal conditioner (Analog devices) |
| 1 | S1 | Electrolytic sensor (Fredericks) |

## Magnetic tiltmeter electronics parts list

| Quantity | Part | Description |
|---|---|---|
| 1 | R1 | 10-k$\Omega$, ¼-W resistor |
| 1 | R2 | 470-k$\Omega$, ¼-W resistor |
| 1 | R3 | 500-k$\Omega$, ¼-W resistor |
| 2 | R4, R6 | 15-k$\Omega$, ¼-W resistor |
| 1 | R5 | 50-k$\Omega$, ¼-W trim pot |
| 1 | U1 | UA741C op amp |
| 1 | AM or DM | 1–10 $V_{dc}$ Analog voltmeter / 1–10 $V_{dc}$ Digital voltmeter |

## PSD Tiltmeter parts list

| Quantity | Part | Description |
|---|---|---|
| 2 | R1, R2 | 10-k$\Omega$–50-k$\Omega$ scaling resistors |
| 9 | R3-R11 | 10-k$\Omega$, 1%, ¼-W resistors |
| 2 | R12, R14 | 5-k$\Omega$, trim pots |
| 2 | R13, R15 | 7.5-k$\Omega$, 1%, ¼-W resistors |
| 3 | R16, R17, R18 | 20-k$\Omega$, ¼-W trim pots |
| 1 | C1 | 4.7-µF, 25-V electrolytic capacitor |
| 3 | U1-U3 | TL082 op amp (T1) |
| 1 | U4 | AD533 (Analog Devices) |
| 1 | M | 1–10 $V_{dc}$ voltmeter |

# Earth-movement sensor

A seismometer is an intriguing device. It's hard to believe that a sensor could be sensitive enough to detect ground vibrations halfway around the earth. A seismometer, or earth-movement sensor, is shown in Fig. 1-68.

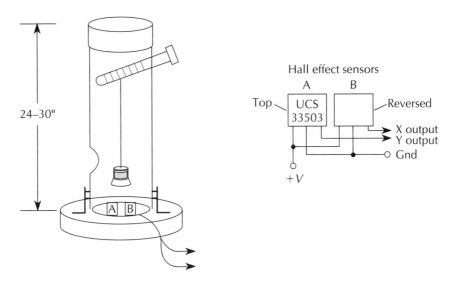

**Fig. 1-68** *Earth-movement sensor.*

The seismometer is an uncomplicated but sensitive magnetic pendulum detector. While this sensor will not pick up earthquakes on the opposite side of the globe, it has been used to pick up nuclear underground testing a few states away. The heart of the pendulum detector is a 3-foot-long × 3-inch diameter PVC tube that houses the pendulum detector. The pendulum must be free to swing unaffected by air currents or movements of the PVC outer tube. The pendulum must be placed in a draft-free, quiet, stable location to operate correctly. The easiest solution to these requirements is a PVC housing mounted to a large round plywood base that is anchored to a concrete slab. Place a cap on the top end of the PVC tube. Next, drill a hole near the cap to accept a long screw that goes through the tube, as shown, to adjust the pendulum height above the detector. Drill a second hole near the bottom of the tube to provide viewing and sensor alignment.

Place a small clear plastic cover over this hole to prevent air currents from affecting the pendulum. Attach a large round plywood base treated with a protective coating, to the PVC. Use three right-angle brackets to attach the tube to the plywood base. Secure the wooden base to a concrete slab to prevent the wooden base and sensor from "dancing" about on the floor in the case of a strong motion event.

The electronics of the earth-movement sensor include two Hall-effect sensors (see Fig. 1-69). Place the two or three-lead Hall sensors at the base of the PVC tube assembly. Place both sensors flat-side down next to each other. Place one face up, i.e., writing up, and flip over the other sensor and connect the leads as shown. Sensor A has a positive output while sensor B produces a negative output with a zero magnetic field. At the center position of the pendulum, the two outputs cancel. These two outputs provide a difference signal, which is further amplified via a TI TLC251 op amp. The gain of the op amp is set for 20. Connect the op-amp's output to a zero-centered voltmeter or bridge circuit. The earth movement will be displayed as a plus and minus swing of the meter.

Next, attach a small magnet to a length of fine string or thread. Wind the free end of the thread around the long screw at the top of the PVC tube. Allow the magnet to hang freely over the Hall sensors at the base of the PVC tube. Adjust the pendulum length so that the magnet is just over but not touching the sensors. Most seismic sensors use a damping scheme to quench excessive swinging movement. This can be accomplished either mechanically or electronically. One approach to damping the pendulum is to place a glass petri dish, filled with mineral oil, on top of the Hall sensors. Adjust the pendulum so that the magnet can swing freely in the mineral oil. The earth-movement sensor is now ready to sense ground movement. This sensor arrangement is very sensitive and, therefore, it should be located in a quiet site away from active living spaces, because it will pick up passing trucks and general nearby movement.

The electronics portion of the earth-movement sensor can also be used as a compass to detect magnetic north. Mount the sensors and the electronics in a small plastic box. Affix a compass rose to the top of the plastic box. Mount the sensors next to each other, perpendicular to the top of the box; i.e., the first side of the sensors are perpendicular to the around. The meter could then be read as the box is rotated. An indication will be given as the box is moved through north, and the compass rose will indi-

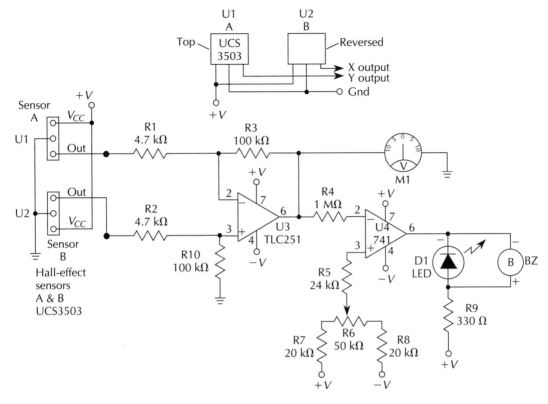

**Fig. 1-69** *Earth-movement-sensor electronics.*

cate the other directions. An LED could be substituted for the meter in this application.

### Earth-movement sensor parts list

| Quantity | Part | Description |
|---|---|---|
| 2 | R1, R2 | 4.7-kΩ, ¼-W resistor |
| 1 | R3, R10 | 100-kΩ, ¼-W resistor |
| 1 | R4 | 1-MΩ, ¼-W resistor |
| 1 | R5 | 24-kΩ, ¼-W resistor |
| 1 | R6 | 50-kΩ ¼-W trim pot |
| 2 | R7, R8 | 20-kΩ, ¼-W resistor |
| 1 | R9 | 330-Ω, ¼-W resistor |
| 1 | D1 | LED |
| 2 | U1, U2 | UCS3503 Hall cells (Allegro Microsystems) |
| 1 | M | 100-0-100 μA meter |
| 1 | BZ | Piezo buzzer |

# New film sensor technology

THIN FILM SENSORS ARE RAPIDLY REVOLUTIONIZING THE FIELD OF sensor technology. The new Piezo film and force/position sensing films are being used in an ever-widening variety of sensing applications and they are beginning to appear in a number of new low-cost sensing products.

## Piezo film sensors

Jacque and Pierre Currie first discovered that quartz crystals produced an electrical charge when deformed. They also discovered that the same crystals changed in dimension when subjected to an electric field. They termed the phenomenon, *piezoelectricity* from the Greek "pressure electricity." One of the first practical applications of the phenomenon was for sonar use in detecting submarines in World War I. In the 1960s, piezoelectric effects were found in organic materials, such as human bones. It was later discovered that polymers, such as polarized homopolymer vinylidene fluoride, developed a highly active piezo response. In the 1970s, Penwalt Corporation developed an exciting new product called *Kynar* or *PVDF*, a piezoelectric film, and a new family of piezo and pyroelectric sensors was born.

The PVDF film can be used in a spectacular array of sensor applications, including magnetic switches, seismic sensors, level sensors, microphones, hydrophones, infrared sensors, optical shutters, and vibration sensors. Kynar film exhibits strong piezo and pyroelectric effects with a broad dynamic range and low acoustic impedance, which makes the PVDF film very attractive for low-cost sensor applications (see Fig. 2-1). The properties of the piezo film heavily depend on the degree and type of crystal structure.

**Fig. 2-1** *Polymer piezolectric sensor.*

PVDF or Kynar film begins with the deformation of crystallites, stretching the extruded PVDF film at temperatures below its melting point, which causes the unit cells to be packed in parallel planes, creating a polar beta phase. Next, the beta-phase polymer must be "polled." Polling exposes the polymer to a high electrical field at elevated temperatures. The level of piezo activity depends upon the polling time, field strength, and temperature. The polling procedure provides a permanent orientation of the molecular dipoles within the polymer, (see Fig. 2-2).

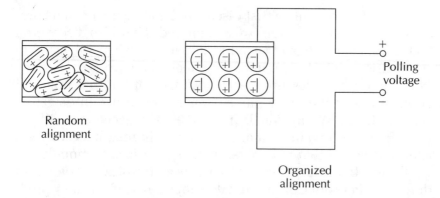

**Fig. 2-2** *Piezo-film process.*

A working voltage applied to the electrodes of the piezo film causes the film to elongate or contract, depending on the polarity of the field. Exposing the film to an alternating field causes both contracting and elongation, one after the other in succession. Conversely, when an external force is applied to the film, the compressive or tensile strain develops a proportional voltage. Note the dual nature of piezoelectricity. The frequency response of the piezo film can vary from 0.005 Hz into the gigahertz region. Tension on the piezo film increases the volume of the material, thus changing the net charge distribution within the film. Under open-circuit conditions, the excess surface charges appear as a voltage difference between the film electrodes. Then, upon closing the circuit, electrons flow to reestablish electrical neutrality, thus producing a usable current. The piezo film requires no external electrical power and, in fact, is a voltage generator and can be used to directly trigger CMOS circuitry when used as a sensor.

The piezo film also acts as a pyroelectric transducer, which can detect thermal radiation. When thermal energy is absorbed, the film expands with the increasing temperature, resulting in a detectable deformation, and a corresponding charge is provided. The reverse effect occurs as the film is cooled. Methods of constructing a pyroelectric detector are shown in Fig. 2-3. A thin semitransparent electrode is used as a front electrode. The radiation passes into the piezo film, which strongly absorbs radiation in the 8–11 km wavelength region. If an infrared-absorbing layer is added to the front electrode, the sensor becomes a broadband absorber of radiation (see Fig. 2-4).

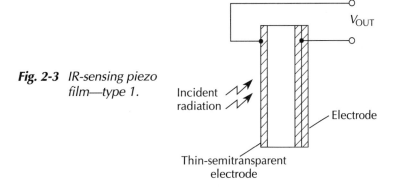

**Fig. 2-3** IR-sensing piezo film—type 1.

Pyroelectric-film sensors are now used extensively and are a preferred alarm sensor among alarm installers, because they generate few false alarms. Pyroelectric or infrared sensors can be made to detect humans and large animals up to 50 feet away, using a low-cost plastic Fresnel lens in front of the film sensor. Coverage can be adjusted to track a long narrow path ranging from 40 to 50 feet, or it can cover a wide area, but with a shortened distance range, depending upon the selected lens.

Piezo film is flexible, lightweight, and tough and is available in a wide variety of thicknesses and configurations. The film pro-

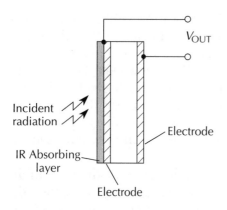

**Fig. 2-4** *IR-sensing piezo film—type 2.*

duces a high output with a wide dynamic range, and it has excellent stability and repeatability at a relatively low cost. However, piezo film has a relatively low "Q," which makes the film unsuitable to very low frequencies when compared to ceramic material. The Kynar film transducers are most efficient at higher frequency ranges, starting at 0.5 to 1 Hz.

A simple flex switch is shown in Fig. 2-5. A rectangular piece of metalized piezo film is bonded to a 5-mil polyester laminate. By flicking or bending the free end of the piezo film, an output is produced. The output is first positive as the polyester is stressed, then becomes negative as the stress is relieved and the film returns to its normal position.

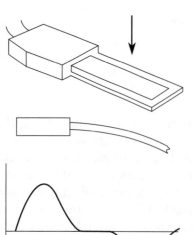

**Fig. 2-5** *Piezo flex switch.*

The film can also be bonded to a strip of spring steel. This type of switch is appropriate when the input force is strong and constant for a period of time. A snap-action switch is shown in Fig. 2-6. Instead of flicking the switch to produce a high impact, you slowly push on the strip, to store up potential mechanical energy. When the strip no longer can withstand the mounting force, output is produced. This type of switch is formed by putting a crease or dimple in the spring-steel strip. This type of switch is also suited to temperature switching. The film is often attached to a bimetal strip, and the different temperature coefficients of the two types of metals flex the switch when heated.

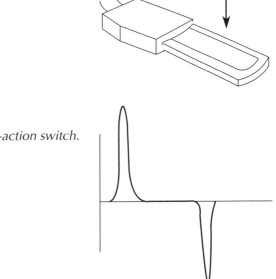

**Fig. 2-6**  *Piezo snap-action switch.*

The piezo film switch can easily be adapted into a magnetic switch by coating a polyester laminate with a ferrous layer or by bonding a metal strip to the polyester piezo film laminate. When a magnet is brought near the switch, the strip will flex, and produce an output (see Fig. 2-7). A variation to the magnetic switch is a sensor capable of measuring displacement velocity and/or acceleration. By using an aluminum strip bonded to a polyester piezo film laminate and then attaching a small weight to the free end of the strip, the device can measure vibration or acceleration.

A "singing switch" type sensor is shown in Fig. 2-8. This type of detector is a novel sensor based on two piezo film strips. One

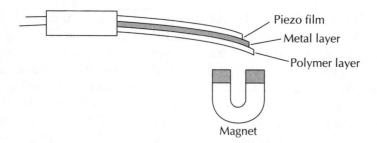

**Fig. 2-7**  *Magnetic piezo switch.*

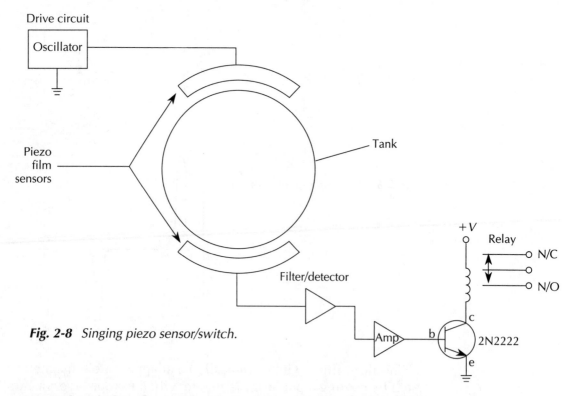

**Fig. 2-8**  *Singing piezo sensor/switch.*

piece is bonded to the inside of a tank or storage vessel, and the other strip is bonded to the outside of the container. One strip is excited with a low-frequency signal source, which is often heard in the audio range (thus, the name, "singing switch"). The second piece of piezo film is made to vibrate at its natural frequency, and it "listens" to the vibrations generated by the source. If the container is pressed (i.e., loaded or filled with a liquid), the frequency is altered and an output is created based on the frequency changes due to the liquid level changes. Picture a series of these piezo film strips arranged vertically inside a large tank or storage vessel. A multiplexing scheme would be used to excite the film strips and interrogate the state of each of the sensor strips in succession.

By arranging a number of small piezo film strips end-to-end, it is possible to build a laser spot follower or precision position sensor. The sensor strips could be scanned and the energized piezo strip could then indicate the precise spot of a laser beam.

A few practical circuit examples employing piezo film are illustrated in Figs. 2-9 through 2-13. The first circuit in Fig. 2-9 depicts an op-amp interface acting as a charge amplifier. This circuit is best suited to small signal applications where low-level vibrations activate the film sensor. The signal level detector in Fig. 2-10, can be used in the presence of large signal-to-noise ratios. This circuit is appropriate for detecting large impacts among low-level vibrations or background noise.

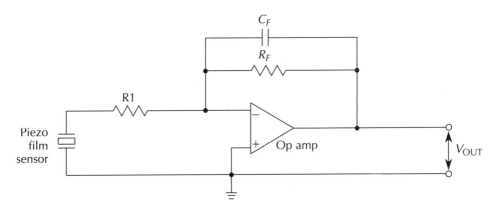

**Fig. 2-9** *Piezo-sensor op-amp amplifier.*

Where the signal-to-noise ratios are low and where impacts or pressures must be discerned from the background, the circuit in Fig. 2-11 is recommended. This circuit uses a common mode-rejection technique based on two sensor switches that are mechanically coupled to cancel the unwanted vibrations that stimulate both sensors. An input or pressure on one switch, but not on the other, produces an output.

Figure 2-12 illustrates a CMOS circuit that senses a single impact or pressure. The flip-flop, once triggered, sounds an audio alarm. CMOS circuitry can be directly driven by the piezo sensor because the signal or voltage output is high level. Figure 2-13, shows a circuit that can sense multiple triggers or impacts and can be used in counting applications.

## 70  New Film Sensor Technology

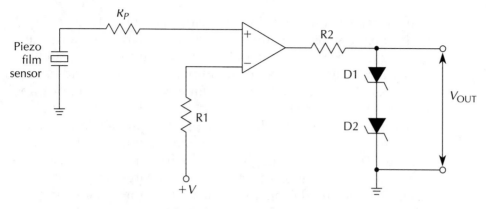

**Fig. 2-10**  *Piezo level detector.*

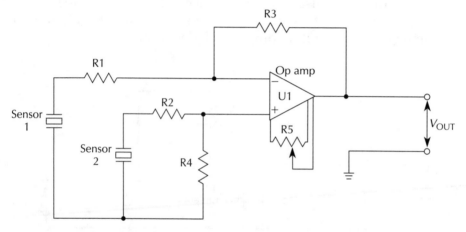

**Fig. 2-11**  *Differential piezo op-amp amplifier.*

The applications for piezoelectric film are limited only by your imagination. The following are some of the more well known applications for Kynar.

- Pushbutton switches
- Keyboards
- Magnetic switches
- Bimetal switches
- Vibration switches
- Accelerometers
- Force
- Level sensors
- Floormat security switches
- Fans
- Laser spot followers
- Microphones
- Hydrophones
- Transducers
- Fence alarms
- Road switches
- Infrared sensors
- Fire/flame sensors
- Optical shutters
- Deformable mirrors
- Speakers

Piezo film sensors 71

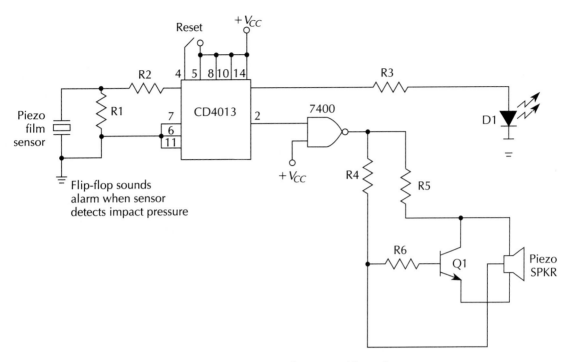

**Fig. 2-12** *Piezo pressure detector with audio output.*

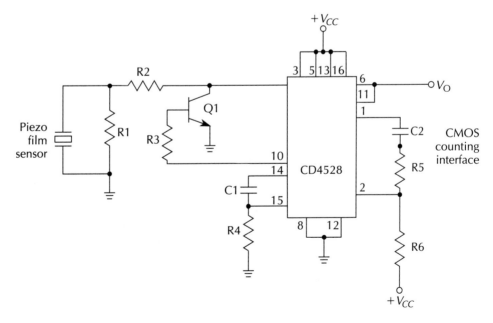

**Fig. 2-13** *Piezo counter interface.*

A sample of Kynar film and application information is available from the Penwalt Corporation, listed in the Appendix. The sample can be configured to demonstrate a number of sensor principles. Penwalt offers many different types of low-cost piezo and pyroelectric sensors, and they can provide a wealth of application assistance.

## Force and position-sensing resistor

The force and position-sensing resistor (FSR) is an emerging new sensor technology with many applications for detecting force and position in one easy operation. Various FSR sensors are shown in Fig. 2-14. The FSR was first developed for musicians who wanted their electronic pianos to play louder as the keys were pressed harder. In the past, there were two principle types of force sensors—piezo polymers and ceramic strain gauges. The fast rise times and high sensitivity of piezo polymers caused too much unwanted acoustical vibration. The new FSR devices are low cost, contain no moving parts, and are impervious to moisture, chemicals, vibration, and magnetism.

FSR devices typically exhibit three decades of dynamic resistance change with respect to force (i.e., 1-k$\Omega$ to 10-M$\Omega$ range). The FSR can be configured into many different types of switch arrays and are ideally suited to keypads, touch switches, alarm sensor mats, pressure sensors, computer pads and tablets, etc. The two basic types of force and position-sensing devices include the FSR-LP linear potentiometer, shown in Fig. 2-15, and the XYZ pad, shown in Fig. 2-16.

Constructing the basic FSR-LP begins with a conductive pattern of interdigiting electrodes deposited on a sheet of polymer. A proprietary conductive polymer is then deposited on a second sheet of mylar. The two sides or sheets of mylar are then faced together so that the conducting fingers are shunted by the conductive polymer. A typical force-VS-resistance curve is shown in Fig. 2-17.

A key element in the basic design of the FSR is the fineness of the conducting fingers. The greater number of shunted fingers produces a greater dynamic range and resolution. A voltage is generally applied between the hot lead and ground, as shown in Figs. 2-18 and 2-19. When a force is applied, the wiper contacts are shunted through the conductive sheet. The voltage output is then read from the wiper and is proportional to the distance along the resistive strip. Thus, a linear potentiometer is created.

Force and position-sensing resistor 73

***Fig. 2-14*** *Force-sensing resistor and resistor/switches.*

***Fig. 2-15*** *Force-sensing resistor.*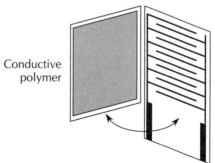

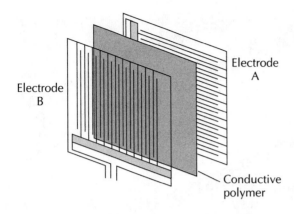

**Fig. 2-16** Force and position sensing resistor pad/tablet.

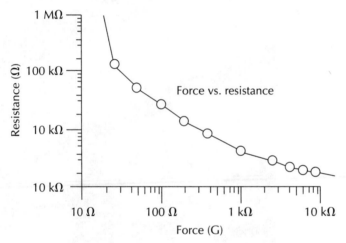

**Fig. 2-17** Force-sensing resistor—force VS resistance curve.

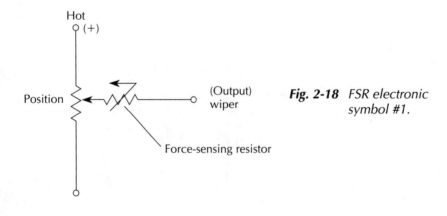

**Fig. 2-18** FSR electronic symbol #1.

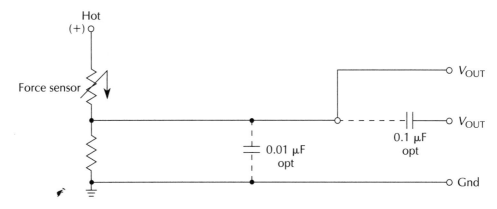

**Fig. 2-19** *FSR electronic symbol #2.*

The contact between the wiper and the resistor itself is momentary; therefore, some sort of sample-and-hold device is needed. Usually, a capacitor between the wiper and ground is often sufficient for this purpose.

The XYZ pad, or tablet, is the second type of force and position-sensing device. Conceptually, two FSR-LPs are set back-to-back (see Fig. 2-16). The position of a force in an x-y coordinate plane, as well as the applied force, can be obtained with the XYZ pad. The XYZ pad is ideal for computer input devices such as tablets, pen stylus pads, and computer mouses.

The XYZ-pad FSR device is capable of a 0.002 inch resolution and a better than 1 percent repeatability accuracy. The XYZ pad is limited to resolving a single force and position measurement at a time and cannot be used for pattern recognition.

A true force sensor gives a constant force independent of the area over which the force is applied. A true pressure sensor gives the same constant force, a reading which is inversely proportional to the area of the applied force. The FSR device in reality lies somewhere in between a true force sensor and a true pressure sensor. A typical FSR device displays a resistance of the square root of the area of the applied force.

The FSR device can be easily interfaced to a wide variety of circuitry. The simple analog interface in Fig. 2-20 displays a force-to-voltage converter, which can be used where a higher current or lower source impedance is necessary. The LF353 or TL071 IC is ideally suited to this interface. The op amp can be used either as a unity-gain buffer or to amplify the FSR's output. Offset current can be trimmed via the 500-Ω trimmer. The gain of the overall system can be set by the ratio of resistors Rf and Rs.

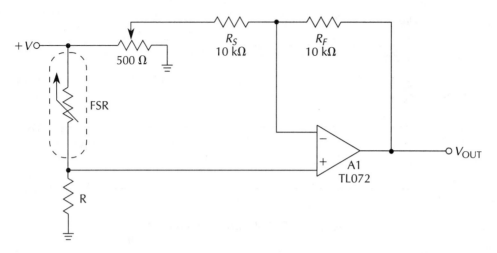

**Fig. 2-20** Force-to-voltage converter.

A force-to-frequency interface is illustrated in Fig. 2-21. An oscillator is formed by the FSR and a Schmitt-trigger IC. At zero pressure, the FSR is an open circuit. Depending on the last stage of the trigger, the output remains either high or low. When the FSR is pressed, the oscillator starts up and the oscillator's frequency is increased or decreased depending on the force applied to the sensor. A 4.7-k$\Omega$ resistor limits current through the FSR. Bypassing this 4.7-k$\Omega$ resistor will cause the curve to steepen at a higher applied force. Connecting a larger value resistor in parallel with capacitor C will quench any tendency to oscillate at very low pressure. Using the oscillator circuit with additional CMOS or TTL circuitry allows you to count activation or strikes applied to the sensor by using leading or trailing edges of the output pulses.

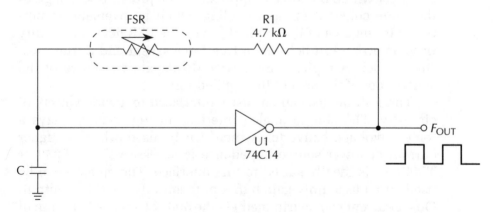

**Fig. 2-21** Force-to-frequency converter.

A few caveats when using the FSR devices include using compression connectors instead of soldering directly to the polymer sensors. You must also ensure that the current through the FSR does not exceed 1 mA/cm$^2$ of footprint activation. Also, FSR devices require a backing material or piece of plastic behind the sensor for consistent activation accuracy. Elastomer overlays can greatly enhance the response characteristics by evenly spreading the activators force.

FSR devices typically respond to pressures between 0.01 to 100 psi, depending on the particular FSR chosen. The FSR sensors are available in a wide variety of sizes and shapes, and many custom devices are available upon request. Table 2-1 lists numerous applications in which the force and position-sensing resistors have been used. A sample FSR device can be obtained by writing on your company letterhead to Interlink Electronics, P.O. Box 40760, Santa Barbara, CA 93103.

**Table 2-1  Force/position-sensor applications.**

| | |
|---|---|
| Graphic pads | Suspension sensors |
| Cursor keys | Accelerometers |
| Mouse devices | Water-level sensors |
| Pen stylus tablets | Panel switches |
| Pressure sensors | Digital tuning controls |
| Piano keyboards | Joysticks |
| Toner-level sensors | Game controllers |
| Foot pedals | Steering wheels |
| Gait-analysis sensors | Gas pedals |
| Artificial-limb force sensors | Robotic touch sensors |
| Dental-bite sensors | Industrial keyboards |
| IV-drip pressure sensor | Deadman safety switches |
| Bed weight-distribution sensor | Smart brake pedals |
| Security tiles/alarm sensors | Oil-pressure sensors |
| Load cells | Limit switches |

# ❖3
# New sensors, ICs, and gas-sensing technology

IN THE PAST TWO DECADES, INTEGRATED CIRCUITS HAVE PLAYED A major role in the wondrous advancements in computing and aerospace electronics. Large-scale integrated circuits now combine a multitude of functions on a single slab of silicon. Engineers, technicians, and hobbyists can now design and build electronic circuits and systems in a much shorter time and at a much lower cost than at any time in the past. The benefits of integrated circuits include speed, low noise, low power consumption, high density, and low cost. Recently, many new sensing and sensor support chips have become available. These low-cost ICs can be used to sense position, motion, speed, liquid level, acceleration, and smoke and gases.

## Piezo accelerometer

We are about to see many new applications for low-cost accelerometers. The new piezoelectric-film accelerometers are now available for less than $10. Traditional accelerometer applications include feedback and control systems, automobile airbag/braking systems, health monitors, security systems, and seismic sensors. Many different types of low-cost accelerometers are currently available. The two major sensor types are the compression sensor and the beam-design-type accelerometers. In the compression design, the inertial mass acts on the entire polymer or film surface. In the beam configuration, the strain is applied to the polymer's cross-sectional area. The beam-type sensors generally provide higher sensitivities.

A general-purpose, low-cost, single-axis compressive sensor is a Penwalt ACH-01, shown in Fig. 3-1. This sensor features

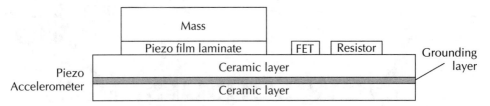

**Fig. 3-1** *Piezo accelerometer.*

wide frequency and dynamic range with good sensitivity. The piezo film produces a linear response from 0.01 to 150 g. A small rectangular mass is mounted over a piezo laminate as shown. The alumina substrate incorporates a low-bias-current field-effect transistor for low output impedance and a high-impedance output load resistor.

Enclose the ACH-01 sensor in an EMI/RFI shielded plastic case with two mounting holes at both ends of the case. The actual usable range of the sensor is between 1 and 25 kHz, with a typical sensitivity of 12 mV/g. The sensor is a three-lead device. A positive voltage is applied to pin 1. The signal output is on pin 2 and ground is applied to pin 3. The power requirement for the ACH-01 ranges from 5 $V_{dc}$ to 24 $V_{dc}$. Couple the output of the accelerometer to an LF363 BI-FET op amp, as shown in Fig. 3-2. Bias the op-amp input with a 180-k$\Omega$ resistor connected to the minus supply. A dual supply is required for this op-amp circuit. Couple the

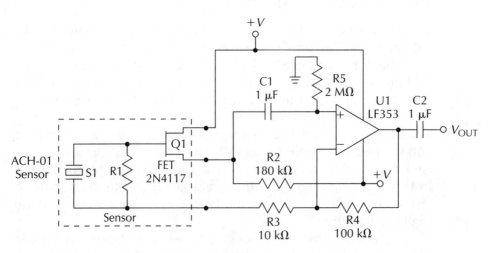

**Fig. 3-2** *Piezo accelerometer interface.*

signal to the LF353 via a 1-µF capacitor connected to the "+" input pin. The op-amp output can be passed onto an optional filter for additional processing as required by your application. The output could be applied to a chart recorder, datalogger, or A/D converter card placed in a personal computer. Other sensors are available, such as the ACH-06-01, which provides up to 2 V/G output, with a frequency range down to 0.01 Hz. The sensitivity of the piezo sensors is generally determined by the film thickness and the amount of inertial mass on the lever arm.

In the future we will see many more applications for these low-cost accelerometers in health systems and in automobile fuel control, braking, and airbag sensor systems.

### Piezo accelerometer interface parts list

| Quantity | Part | Description |
|---|---|---|
| 1 | R1 | (Internal) HiZ resistor |
| 1 | R2 | 180-k$\Omega$, ¼-W resistor |
| 1 | R3 | 10-k$\Omega$, ¼-W resistor |
| 1 | R4 | 100-k$\Omega$, ¼-W resistor |
| 1 | R5 | 2-$\Omega$, ¼-W resistor |
| 1 | C1, C2 | 1-µF, 25-V capacitor |
| 1 | Q1 | (Internal) FET |
| 1 | U1 | LF353 op amp |
| 1 | S1 | ACH-01 piezo sensor (Atochem/Penwalt) |

## Optical transceiver

The complete optical transceiver shown in Fig. 3-3 is a new offering from Cherry Semiconductor Corporation. The bipolar LSI CS258 chip can be used in intrusion alarms, spot alarms, window/door sensors, particle/dust alarms, and industrial parts counters. An infrared LED, phototransistors, a few discrete components, and the CS258A form a complete optical transceiver that can be used as a stand-alone alarm system or as a sensor module wired to a central alarm panel. A sensor module can be created on a small printed circuit board measuring 1½×2¼ inches. The wiring from each module can be daisy-chained back to the central alarm panel.

The CS258A transceiver chip is a complete system that incorporates a three-pole filter network with a gated pulse-detection scheme that prevents false alarm conditions. When the infrared (IR) source is blocked from the detectors, no output is

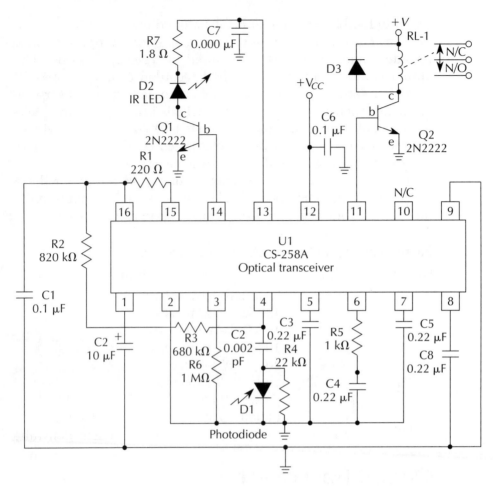

**Fig. 3-3** *Optical transceiver chip.*

present. However, as soon as the IR beam is detected by the phototransistor, (i.e., an object is moved from the light path) an output is presented on pin 11. The output can drive a low-current relay, which, in turn can activate a siren or bell.

The CS258A transceiver can be operated from a 5-V power source and has a maximum voltage rating of 7 $V_{dc}$. The external LED drive transistor in this circuit example boosts the current of D2 to increase the range between the LED and the phototransistor. The circuit is straightforward; however, pay particular attention to the dual ground system. A high-level ground on pin 9 is separate from the low-level ground on pin 2. The optical transceiver can be used in a wide range of alarm applications.

## Optical transceiver parts list

| Quantity | Part | Description |
|---|---|---|
| 1 | R1 | 220-Ω, ¼-W resistor |
| 1 | R2 | 820-Ω, ¼-W resistor |
| 1 | R3 | 680-kΩ, ¼-W resistor |
| 1 | R4 | 22-kΩ, ¼-W resistor |
| 1 | R5 | 1-kΩ, ¼-W resistor |
| 1 | R6 | 1-Ω, ¼-W resistor |
| 1 | R7 | 1.8-Ω, ½-W resistor |
| 1 | C1 | 0.1-µF, 25-V capacitor |
| 1 | C2, C6 | 0.002-µF, 25-V capacitor |
| 1 | C3, C4, C5, C8 | 0.22-µF, 25-V capacitor |
| 1 | C7 | 1000-µF, 25-V electrolytic |
| 1 | D1 | Photodiode detector or optocoupler |
| 1 | D2 | IR LED or optocoupler |
| 1 | D3 | 1N4001 silicon diode |
| 1 | Q1 | 2N2222 pnp transistor |
| 1 | U1 | CS258A (Cherry Semiconductor) |
| 1 | RL-1 | 6–12-V SPST relay |

# Proximity detector

The new Cherry Semiconductor CS209 bipolar *proximity detector*, shown in Fig. 3-4, includes an oscillator, demodulator, level detector, output stage, and regulator. The proximity detector is versatile and can be used in counting and speed-detection applications, coin detectors, burglar alarm switches, and metal detectors. The proximity detector features an LC tank oscillator between pins 2 and 3.

To experiment with the CS209 proximity detector, purchase or wind a 100-µH coil on a ferrite core. A 50–70-turn coil of 26-gauge enameled wire is a good beginning. Once the coil has been measured, that value can be "plugged" into the formula to obtain an optimum capacitor value. The "Q" of the LC oscillator is very important in obtaining resonance and, ultimately, the maximum range of the detector.

A resonant frequency of 200–600 Hz is recommended for best results. The frequency of the tank circuit is half the square root of capacitance (C) times inductance (L). A frequency counter can be connected between pins 3 and 8 to measure the actual frequency. The maximum range can be fine tuned by adjusting or tweaking the value of the rf resistor between pins 1 and 8.

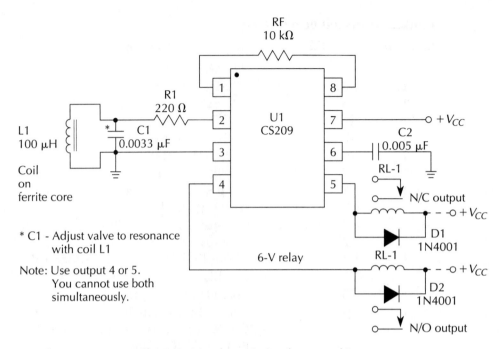

**Fig. 3-4** *Metal proximity-detector chip.*

The CS209 detector has two active high-level outputs capable of driving a low-current miniature relay. Outputs are available on pins 4 and 5. The output on pin 4 is a normally open output and pin 5 is a normally closed output. Select either output and then connect a low-current relay between the selected output and the positive power-supply voltage. The output is TTL-level compatible and can drive logic circuits. The CS209 is a very compact eight-pin miniature DIP integrated circuit with a minimum of external components. The chip is quite versatile and can be used in many alarm sensing applications. Cherry Semiconductor provides samples and application information on request.

## Metal proximity-detector parts list

| Quantity | Part | Description |
|---|---|---|
| 1 | R1 | 220-Ω, ¼-W resistor |
| 1 | R2 | 10-kΩ, ¼-W resistor |
| 1 | C1 | 0.0033-µF, 25-V adj for resonance |
| 1 | C2 | 0.005-µF, 25-V capacitor (disk) |
| 1 | L1 | 100-µH coil on ferrite core |
| 1 | D1 | 1N4001 silicon diode |
| 1 | U1 | CS209 (Cherry Semiconductor) |
| 1 | RL-1 | 5-V miniature relay |

# Smoke detector

A compact photoelectric smoke-detection system can be constructed using the Cherry Semiconductor CS235 integrated circuit. A pulsed infrared LED and a silicon photodiode are used together as the light source and detector pair. With the addition of a few discrete components, you can create a sensitive and reliable smoke detector. By cascading a number of these detector modules, a complete house smoke-detector system can be fabricated.

The detector system consumes very little power. It pulses the system once every 10 seconds for 20 milliseconds. During the second half of the 20-ms period, the LED is pulsed and the detector samples for a smoke condition. After the first alarm signal is detected, a sample rate increases to a 2-second interval. After three consecutive alarm level samples are accumulated, the logic signal drives the output latch on pin 3. This synchronous detection method has very high noise rejection and provides excellent reliability.

The internal latch is capable of sinking 100-mA of current and will drive a low-current relay directly. The CS235 operates from a 12-$V_{dc}$ supply. The latch current is determined by resistor RL, as shown in Fig. 3-5, and depends upon the relay that you choose in your particular system.

The smoke-detector system uses a two-ground system. A high-level ground is present on pin 4 and the low-level ground is connected on the voltage divider on pin 5. The CS235 smoke detector system is easy to construct and will provide a reliable protection system. The smoke detector can be operated on a 12-V battery trickled-charged from a 110-V source.

## Photoelectric smoke detector parts list

| Quantity | Part | Description |
|---|---|---|
| 1 | R1 | 20 MΩ, ¼-W resistor |
| 1 | R2 | 75-kΩ, ¼-W resistor |
| 1 | R3 | 6-kΩ, ¼-W resistor |
| 1 | R4 | 510-Ω, ¼-W resistor |
| 1 | R5 | 100-kΩ, trim pot |
| 1 | R6 | 18-kΩ, ¼-W resistor |
| 1 | R7 | 12-MΩ, ¼-W resistor |
| 1 | R8 | 36-Ω, ¼-W resistor |
| 1 | R9 | 270-kΩ, ¼-W resistor |
| 2 | R10, R11 | 150-kΩ, ¼-W resistor |

| Quantity | Part   | Description                         |
|----------|--------|-------------------------------------|
| 1        | RL     | Current limiter adjusted for 100 mA |
| 1        | C1     | 0.47-µF, 25-V capacitor             |
| 1        | C2     | 150-µF, 25-V capacitor              |
| 1        | C3     | 0.0068-µF, 25-V capacitor (disk)    |
| 2        | C4, C5 | 0.47-µF, 25-V capacitor             |
| 2        | C6, C7 | 0.001-µF, 25-V capacitor (disk)     |
| 1        | D1     | Status LED                          |
| 1        | D3     | LED                                 |
| 1        | U1     | CS235 (Cherry Semiconductor)        |
| 1        | RL-1   | 9–12-$V_{dc}$ SPST relay            |

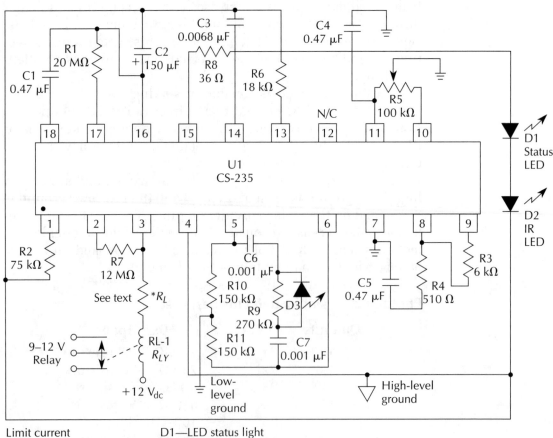

*Fig. 3-5  Photoelectric smoke-detector chip.*

# Fluid detector

The National LM1830 fluid detector detects the presence, absence, or level of a polar liquid, such as water. The LM1830 is a linear bipolar integrated circuit that operates from a wide range of operating voltages. The fluid detector shown in Fig. 3-6 determines the presence of a fluid by comparing the resistance of the fluid between the probes and the internal resistance of the IC. An ac signal is used to overcome the plating problems associated with a dc voltage source. Provisions are available for connecting an external resistance in applications where the fluid impedance is of a different magnitude than an internal resistor in the IC. As the probe resistance increases above a preset value, the internal capacitor produces a signal.

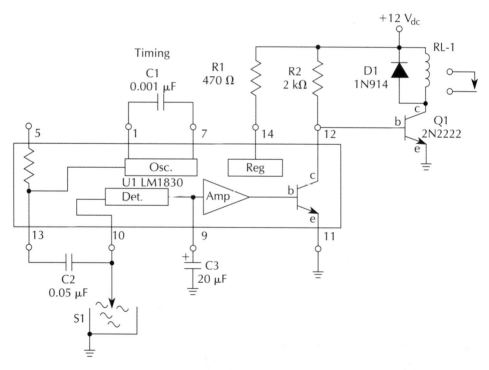

**Fig. 3-6** *Fluid-detector switch.*

The internal oscillator is coupled to the base of an open collector transistor so that the output signal can drive an LED, a speaker, or low-current relay. A 0.001-μF timing capacitor between pins 1 and 7 produces a 6-kHz signal on pin 5 of the fluid detector. A filter capacitor on pin 9 enables the fluid detector to operate at a constant output. Removing this filter allows a 50% duty cycle. When used with a speaker, the output will cycle on and off.

Although the LM1830 detector was primarily designed for sensing conductive liquids, it can also be used in the direct-coupled mode. A variable resistance sensing device, such as a light-dependent resistor or a thermistor, could be used as the sensing device, as shown in Fig. 3-7. The fluid detector is a complete, low-cost level-detector system that can be used in many different sensing applications. The detector can drive a low-current relay as a local alarm. The relay contacts could be wired to other sensors, and the multiple sensors could activate a central alarm panel.

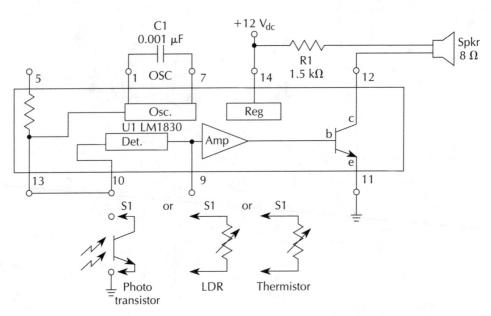

**Fig. 3-7** *Fluid detector with sound output.*

## Fluid-detector switch parts list

| Quantity | Part | Description |
|---|---|---|
| 1 | R1 | 470-Ω, ¼-W resistor |
| 1 | R2 | 2-kΩ, ¼-W resistor |
| 1 | C1 | 0.001-µF, 25-V capacitor (disk) |
| 1 | C2 | 0.05-µF, 25-V capacitor (disk) |
| 1 | C3 | 20-µF, 25-V electrolytic capacitor |
| 1 | D1 | 1N914 silicon diode |

| Quantity | Part | Description |
|---|---|---|
| 1 | Q1 | 2N2222 pnp transistor |
| 1 | U1 | LM1830 (National Semiconductor) |
| 1 | S1 | Metal container with polar liquid |
| 1 | RL-1 | 9–12-V SPST relay |

**Fluid detector with sound output parts list**

| Quantity | Part | Description |
|---|---|---|
| 1 | R1 | 1.5-k$\Omega$, ¼-W resistor |
| 1 | C1 | 0.001-µF, 25-V capacitor (disk) |
| 1 | U1 | LM1830 (National Semiconductor) |
| 1 | S1 | MRD 300 phototransistor or thermistor |
| 1 | SPKR | 8-$\Omega$ speaker |

# Over/under current detector

The LM1946 over/under current-limit sensor from National Semiconductor is a set of five independent comparators with a voltage reference supply that can be used as over/under current limit detectors, or two comparators can be wired together to form a window detector. The versatility of the LM1946 enables the IC to be used in industrial or automotive system applications.

The current is sensed by monitoring the voltage drop across a wiring harness or a PC board trace, or by using an external sense resistor that feeds the load. Provisions are made to compensate for power-supply variations, wiring-harness variations, and temperature changes. Once a limit is reached, a comparator turns on its output latch, which can drive an LED or low-current relay. Additionally, one side of the load can be grounded, which is important in automotive applications.

Resistors R1 and R2 set the comparator's threshold voltage range. Values of sense voltages greater than 100 mV to the comparator create an off or low output. Sense voltages less than 100 mV turns the output to an on or high state. The comparator is only turned on or activated when inputs are above ground by at least 3 V.

Figure 3-8 depicts a typical load circuit. S1 represents a switch or sensor. The sense resistor $R^S$ can be an actual resistor, such as a 0.1-$\Omega$ carbon resistor, or it can be represented by an auto wiring harness. The load circuit could be an autolamp or a motor, and the LM1946 could be used as a motor stall detector. The LM1946 is quite flexible and can be configured to many autoalarm installations, as well as industrial sensing applications.

## 90 New Sensors, ICs, and Gas-Sensing Technology

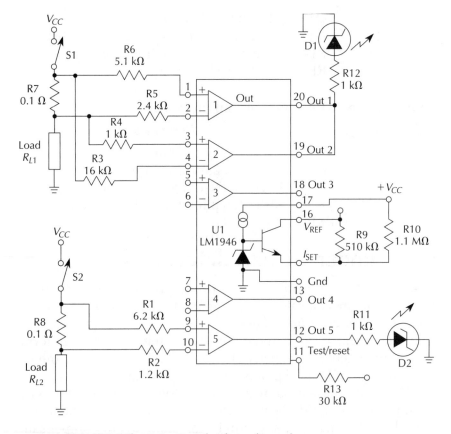

**Fig. 3-8** Over/under voltage detector.

## Over/under voltage detector parts list

| Quantity | Part | Description |
|---|---|---|
| 1 | R1 | 6.2-k$\Omega$, ½-W resistor |
| 1 | R2 | 1.2-k$\Omega$, ½-W resistor |
| 1 | R3 | 16-k$\Omega$, ½-W resistor |
| 1 | R4 | 1-k$\Omega$, ½-W resistor |
| 1 | R5 | 2.4-k$\Omega$, ¼-W resistor |
| 1 | R6 | 5.1-k$\Omega$, ¼-W resistor |
| 2 | R7, R8 | 0.1-$\Omega$, ½-W resistor |
| 1 | R9 | 510-k$\Omega$, ¼-W resistor |
| 1 | R10 | 1.1-M$\Omega$, ¼-W resistor |
| 2 | R11, R12 | 1-k$\Omega$, ¼-W resistor |
| 1 | R13 | 30-k$\Omega$, ¼-W resistor |
| 2 | RL-1, RL-2 | Load device lamp, motor, etc. |
| 2 | D1, D2 | Red LED |
| 1 | U1 | LM1946 (National Semiconductor) |

# Tachometer/speed detector

The LM2907 tachometer/speed detector is a handy device for measuring wind/auto/boat speed, position sensing, remote control systems, and, more recently, antiskid braking systems. The tachometer/speed detector consists of an input amplifier, a charge-pump frequency-to-voltage converter, an op-amp comparator, and an uncommitted open-collector transistor, which can be used to drive relays, motors, analog meters, and computer data loggers. The LM2907/2917 series chips are single-supply voltage building blocks, available in both 8- and 14-pin packages. The 14-pin package provides a differential-input option, which allows the user the option of tailoring his own particular isolation and input preferences. The LM2917 chip also includes an additional internal zener diode for regulation.

Operation of the tachometer/speed detector begins with the input op amp, which provides a hysteresis switching input level of ±15 mV, where noise may be present on the input signal. Total noise rejection is allowed below this amplitude if there is no actual signal. A low-frequency signal is applied to the input of the charge pump on pin 1. The emitter output on pin 4 is connected to the inverting input of the op amp so that pin 4 follows pin 3 and provides a low-impedance output voltage proportional to the input frequency. Linearity of the output is better than 0.3%. The capacitance of C1 provides an internal compensation for the charge pump and should be larger than 100 pF. The type of capacitor used for timing is very important to the accuracy of the system.

Figure 3-9 illustrates a magnetic variable-reluctance sensor used with an analog-voltmeter speed indicator. The input to the tachometer/speed detector can take many forms. The diagram in Fig. 3-10 shows an ac-coupled input, a bandpass-filtered input, and a differential-input circuit. A variable-capacitor sensor in Fig. 3-11, could also be used as the input sensing device. The capacitive sensor is connected at pin 2 and ground. Pin 1 is connected to the junction of two resistors as shown, a 5-kΩ resistor to ground and a 500-kΩ resistor coupled to a 110-$V_{ac}$ 60-Hz voltage source. The output of the tachometer/speed detector can be configured in different ways, depending on your particular application. The output could be used to drive relays, meters, analog/digital meters, A/D cards in computers.

The computer interface depicted in Fig. 3-12 can perform data-logging operations directly to a personal computer. To start a conversion cycle, the processor generates a reset pulse to dis-

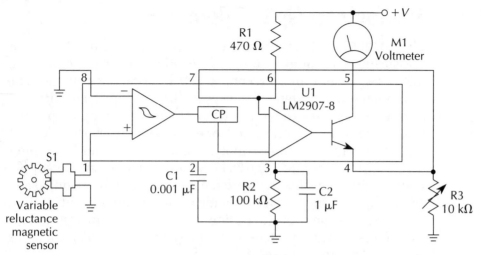

**Fig. 3-9** Tachometer.

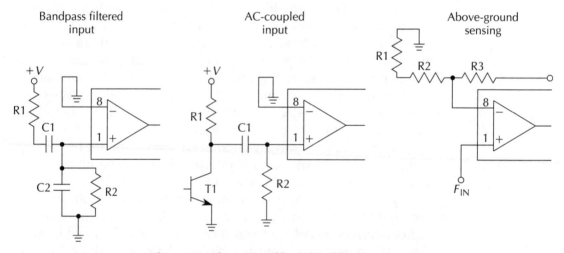

**Fig. 3-10** Alternative filtered tachometer inputs.

charge the integrating capacitor C2. Each complete clock cycle generates a charge and discharge cycle on C1. This results in two steps per cycle being added to C2. As the voltage on C2 increases, clock pulses are returned to the processor. When the voltage on C2 steps above the analog input voltage, the data line is clamped and C2 ceases to charge. The processor then counts the number of clock pulses received after the reset pulse and is loaded with a digital number, which is representative of the input voltage. By making $C^2/C^1=1024$, an 8-bit A/D computer interface can be easily constructed.

Position-sensitive detector 93

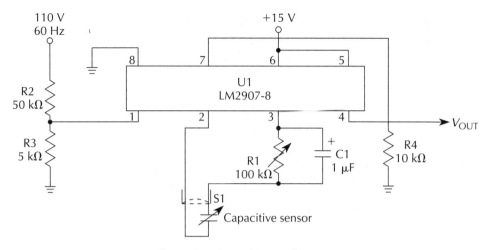

**Fig. 3-11** *Capacitive tachometer.*

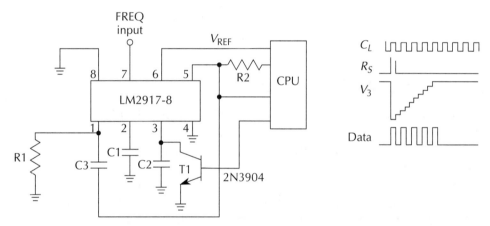

**Fig. 3-12** *Tachometer with computer interface.*

The LM2907 tachometer/speed detector chips can greatly simplify your system design when you need to sense speed or position detection. A simple mechanical propeller can drive a slotted metal wheel mounted in close proximity to the S sensor and can be adapted to many speed sensing applications. You can easily create a wind or marine speed indicator with a handful of inexpensive components.

## Tachometer parts list

| Quantity | Part | Description |
|---|---|---|
| 1 | R1 | 470-Ω, ¼-W resistor |
| 1 | R2 | 100-kΩ, ¼-W resistor |

| Quantity | Part | Description |
|---|---|---|
| 1 | R3 | 10-kΩ potentiometer |
| 1 | C1 | 0.001-µF, 25-V capacitor (disk) |
| 1 | C2 | 1-µF, 25-V capacitor |
| 1 | U1 | LM2907-8 (National Semiconductor) |
| 1 | S1 | Variable reluctance sensor |
| 1 | V | 0–10 V panel meter |

**Capacitive tachometer parts list**

| Quantity | Part | Description |
|---|---|---|
| 1 | R1 | 100-kΩ potentiometer |
| 1 | R2 | 50-kΩ, ½-W resistor |
| 1 | R3 | 5-kΩ, ½-W resistor |
| 1 | R4 | 10-kΩ, ¼-W resistor |
| 1 | C1 | 1-µF, 25-V capacitor |
| 1 | U1 | LM2907-8 (National Semiconductor) |
| 1 | S1 | Capacitive-type sensor |

# Position-sensitive detector

The silicon position-sensitive detector (PSD), shown in Fig. 3-13 is a relatively new optoelectronic sensor that can provide continuous position data of light traveling over its sensitive surface. The PSD can be used for optical position and angle sensing, laser spot following, remote control systems, and cosmic particle detectors, as well as displacement and vibration sensors and auto range finders. Major features of the PSD include high position resolution, wide spectral response, high speed, and good reliability.

The three-lead silicon PSD consists of monolithic pin photodiodes with either one or two uniform resistive surfaces. The

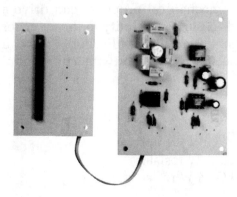

*Fig. 3-13* Position-sensitive detector (PSD) and interface.

PSDs come in many different configurations and sizes including one or two dimensional outputs for x and y position indication.

Figure 3-14 shows the structure of a PSD device. The PSD consists of three separate layers; first is the p-layer at one end and an n-layer at the opposite end with an I or current layer in between. Light striking the PSD is converted photoelectrically and detected by the two electrodes at the p-layer. When a spot of light falls on the PSD, an electric charge proportional to the light energy is generated at the incident position. The electric charge is driven through the resistive p-layer and is collected by the electrodes. Because the resistivity of the p-layer is uniform, the photocurrent collected by an electrode is inversely proportional to the distance between the light-spot position and the electrode.

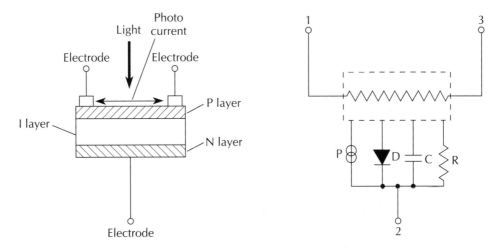

*Fig. 3-14* PSD diagram.

The PSD shown is a one-dimensional type. However, two-layer, two-directional PSDs are available that will display x and y position. Essentially, the x-y device is a two-layer device with the addition of a dc restoration device added between the final op amp U5 and the analog divider.

The circuit shown in Fig. 3-15 is a one-dimensional position device. A bias voltage is applied to the center lead of the PSD with a 4.7-μF capacitor to ground. The two active electrodes are directed to a linear op amp at LF351. Resistors R1 through R9 are 10 kΩ. The two rf resistors on the op-amp inputs are used for range or level adjustment. All resistors should be 1% to give best results. The output of the final op amp (U5) is directed to an analog divider, such as an Analog Devices, AD533. The output of the

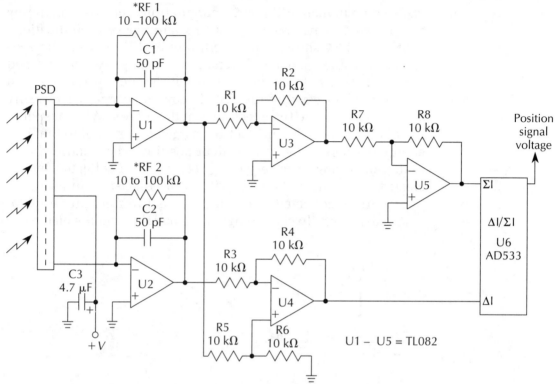

**Fig. 3-15** *PSD interface.*

divider is an analog voltage, ±10 V as selected by the user. The position-voltage output can be coupled to an A/D card in a personal computer as a single-channel input. Another approach is to eliminate the voltage divider and separately connect both the ΔI and ΣI outputs as two channels into an A/D card. This requires the computer to do a bit more computation, but isn't that what computers are for? The PSD is quite easy to implement and works very well for precise position indication.

## PSD interface parts list

| Quantity | Part | Description |
|---|---|---|
| 8 | R1–R8 | 10-kΩ, 1%, ¼-W resistor |
| 2 | RF1, RF2 | 1–100-kΩ, ¼-W scaling resistors |
| 5 | U1-U5 | TL082 (Texas Instruments) |
| 1 | U6 | AD533 (Analog Devices) |
| 1 | PSD | PSD detector (Hamamatsu, Inc.) |

# Twilight sensor

The ULN3390T optoelectronic twilight sensor is a new chip from Allegro Semiconductor, formerly Sprague Electric. The twilight sensor can be used for emergency and outdoor/indoor lighting applications. The ULN3390T is a monolithic IC containing a photodiode, low-level amplifier, comparator, voltage regulator, and output driver, as shown in Fig. 3-16. The comparator is fabricated to give the sensor a built-in hysteresis value of 50%. The ULN3390T has temperature-compensated trip points as well as protection against bright light damage. The ULN3390T opto switch is a significant improvement over previous opto devices, because the IC is much more stable over time and temperature than cadmium sulfide cells, and it requires fewer components and no calibration. The ULN3390T opto switch typically turns on as illumination falls below 10 µW/cm². The switch points can be factory adjusted per customer specifications.

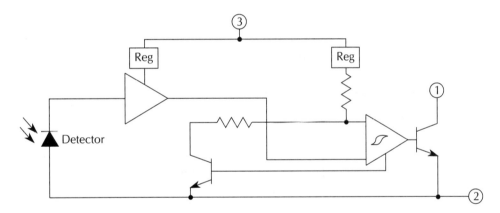

*Fig. 3-16* Optical twilight sensor—block diagram.

The twilight switch is very easy to implement and is shown in Fig. 3-17. The device ratings of the ULN3390T allow up to 25 $V_{dc}$ as a power source, and maximum current output is 25 mA. The circuit is shown powered by a 12-$V_{dc}$ supply. An external transistor drives a low-current relay, which can activate a dc lamp, or the relay shown can activate a higher current relay, which could switch 110 $V_{ac}$ to a lamp or motor, if desired. The twilight sensor could be used to turn on indoor lamps as darkness falls, to fool would-be burglars into believing someone is at home, or the ULN3390T could be used to activate low-voltage

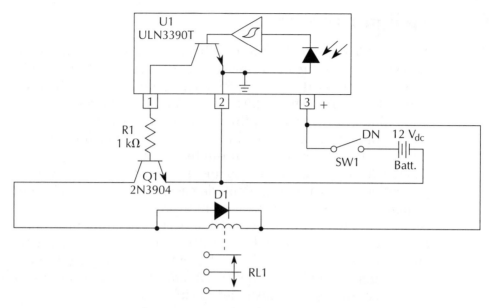

*Fig. 3-17* Optical twilight switch sensor.

driveway or walkway lamps. Call or write Allegro Semiconductor for their new catalog of optoelectric and Hall-effect devices.

## Opto-twilight sensor switch parts list

| Quantity | Part | Description |
| --- | --- | --- |
| 1 | R1 | 1-k$\Omega$, ½-W resistor |
| 1 | D1 | 1N4002 silicon diode |
| 1 | Q1 | 2N3904 transistor |
| 1 | RL1 | 9–12-V SPST relay |
| 1 | SW1 | SPST toggle switch |
| 1 | BATT | 9–12-V battery |

# Multiplexed Hall-effect sensor

The Allegro UGN3055U multiplexed two-wire Hall-effect sensor is an exciting new digital magnetic-sensing IC, which is capable of communicating over a two-wire power/signal bus. By using a sequential addressing scheme, the device responds to a signal on the bus and returns the diagnostic status of the IC, as well as the status of each monitored external magnetic field. As many as 30 sensors can function on the same two-wire bus, and the manufacturer projects up to 60 sensors could be implemented in the future.

A functional diagram of the sensor is shown in Fig. 3-18. This integrated circuit is ideal for multiple-sensor applications where minimization of the wiring harness is desirable. The unique UGN3055U sensor is available in two temperature ranges, −20°C to +85°C and −40°C to +125°C. Alternate magnetic and temperature specifications are available upon request. The UGN3055U sensor is ideal for commercial/industrial applications as well as home alarm systems. The sensor can be mounted on a door or window jam to sense a small magnet recessed in a door or window, or the sensor could be used with an external sensor or switch attached to pin 3 of the IC, as shown in Fig. 3-19. Each sensor has a factory-specified predefined address. At present, 30 addresses are available from a maximum of 60.

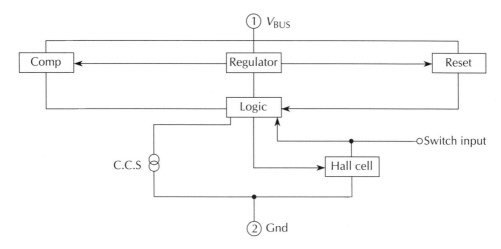

*Fig. 3-18* *Multiplexed Hall sensor block diagram.*

The device can be addressed by modulating the supply voltage, as shown in Fig. 3-20. A preferred addressing protocol is as follows: the bus supply voltage is first brought down to 0 V, so that all devices on the bus may be reset. The voltage is then raised to the address low voltage, $V_L$, and the bus quiescent current is then measured. The bus is then toggled between $V_L$ and $V_h$, with each positive transition representing an increment in the address bus. After each voltage transition, the bus current is monitored for diagnostic and signal responses from the UGN3055U sensor. When a sensor detects a bus address equal to its factory-programmed address, it responds with an increase in supply current drain ($I_s$). This response is during the high portion of the address cycle. This *diagnostic response*

100  New Sensors, ICs, and Gas-Sensing Technology

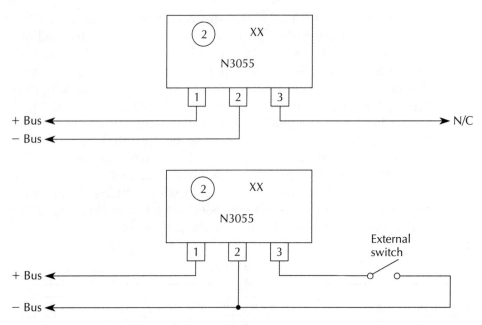

**Fig. 3-19**  Multiplexed Hall sensor.

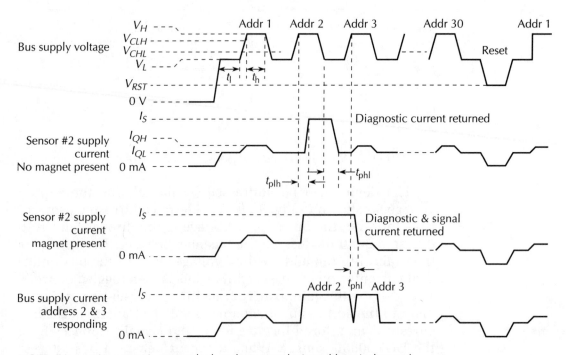

NOTE: Diagnostic current is returned when the preset device address is detected.
Signal current is returned when the correct address and magnetic field are both detected.

**Fig. 3-20**  Bus-timing diagram.

is used as an indication that the sensor is alive and well. If the sensor detects an ambient magnetic field, it also responds with $I_s$ during the low portion of the address cycle.

This response is called the *signal response* when the next positive transition is detected. The sensor becomes disabled, and its contribution to the bus signal current returns to $I_q$.

Figure 3-20 displays the previously described addressing protocol. The top trace represents the bus-voltage transitions as controlled by the bus driver. The second trace represents the bus-current contribution of the sensor (address 2). The diagnostic response from the sensor indicates that it detected its address on the bus. However, no signal-response current is returned, which indicates that a sufficient magnetic field is not detected at the chip surface. The third trace represents the current drain of sensor 2 when a magnetic field is detected. The last trace represents the overall bus current drain when sensors 2 and 3 are present. While sensor 2 returns a diagnostic and signal current, sensor 3 only returns a diagnostic current.

The UGN3055U Hall sensor has been designed to respond to an external magnetic field greater than $B_{op}$. This is accomplished by amplifying the output of an on-chip Hall transducer and feeding it into a threshold detector. In order that the bus current is kept to a minimum, the transducer is kept powered down until the sensor is addressed. The UGN3055U can also detect the status of an external switch, as shown in Fig. 3-19.

The bus should be controlled by a microprocessor for several reasons. First, the sensor address information may be stored in ROM in the form of a lookup table. Second, bus faults can be pinpointed by the microprocessor by comparing the diagnostic response to the expected response in the ROM lookup table. Third, the microprocessor, along with an A/D converter, can self-calibrate the quiescent currents in the bus and thus be able to detect a signal response easily. The microprocessor can also be used to filter out random line noise by digitally filtering the bus response. The microprocessor can easily keep track of the signal responses and initiate the proper action, i.e., light a lamp and sound an alarm. Optimally, the microprocessor could control bus-driving circuitry that will accept TTL level inputs to drive the bus, and will return an analog-voltage representation of the bus current.

The bus driver can be easily designed using a few op amps and resistors and a couple of transistors. Figure 3-20 illustrates a schematic of a suitable bus driver that is capable of providing the 6- to 9-V transitions, thus resetting the bus and providing an analog measurement of the current for use by the A/D input of a microprocessor. The address pin provides a TTL-compatible input to control the bus supply. A high (+5 V) switches Q1 on and sets the bus voltage to 6 V through the resistor divider of R4, R5, and the zener diode. A low input switches Q1 off and sets the bus voltage to 9 V. This voltage is fed into the positive input of the op amp OP-1 and is buffered and made available at the bus supply.

The bus reset control is also available in the form of a TTL-compatible input. When this input, which is marked "RESET," goes high, Q2 is switched on and the positive input of the op amp is set to the saturation voltage of the transistor, resetting the bus. The processing of the bus current available at the A/D input pin of Fig. 3-21 is best accomplished by feeding it into the A/D input of a microprocessor. If the flexibility of a microprocessor is not available, then this signal could be fed into a threshold-detection circuit, and the output could drive a display device. A complete system diagram of a multiplexed Hall-effect alarm system is shown in Fig. 3-22, which can be used with up to 30 stations. This intriguing integrated circuit is described in more

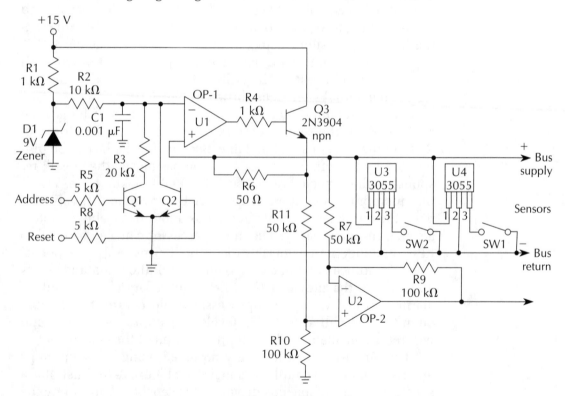

**Fig. 3-21** *Multiplexed Hall-sensor bus interface.*

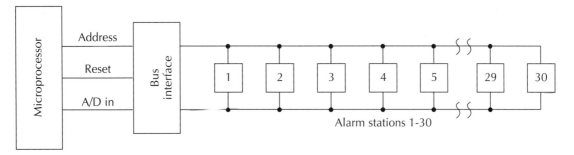

***Fig. 3-22*** *Multiplexed Hall-sensor system diagram.*

detail along with many other Hall-effect devices in the new Allegro AMS500 catalog.

**Multiplexed Hall sensor bus interface parts list**

| Quantity | Part | Description |
|---|---|---|
| 2 | R1, R4 | 1-kΩ, ¼-W resistor |
| 1 | R2 | 10-kΩ, ¼-W resistor |
| 1 | R3 | 20-kΩ, ¼-W resistor |
| 1 | R5 | 5-kΩ, ¼-W resistor |
| 1 | R6 | 50-Ω, ¼-W resistor |
| 3 | R7, R8, R11 | 50-kΩ, ¼-W resistor |
| 2 | R9, R10 | 100-kΩ, ¼-W resistor |
| 1 | C1 | 0.001-μF 25-V capacitor (disk) |
| 1 | D1 | 9-V zener diode (NTE139A) |
| 3 | Q1, Q2, Q3 | 2N3904 transistor |
| 1 | U1 | OP-1 op amp (Analog Devices) |
| 1 | U2 | OP-2 op amp (Analog Devices) |
| 2+ | U3, U4, Uxx | UGN3055U Hall-effect sensors |
| 2+ | SW-1, SW-2, SW... | SPST sensors or switches |

# Videophone

The Vidram 027 videophone chip is one of the most exciting integrated circuits to come down the silicon alley. The 48-pin LSI chip (see Fig. 3-23) can send high-resolution still video pictures down a twisted pair or through the public telephone system. The picturephone also has some other interesting applications, including remote video monitoring in alarm/surveillance systems, as well as amateur-radio applications, which would allow sending pictures via radio waves. The picturephone chip is available

*104* New Sensors, ICs, and Gas-Sensing Technology

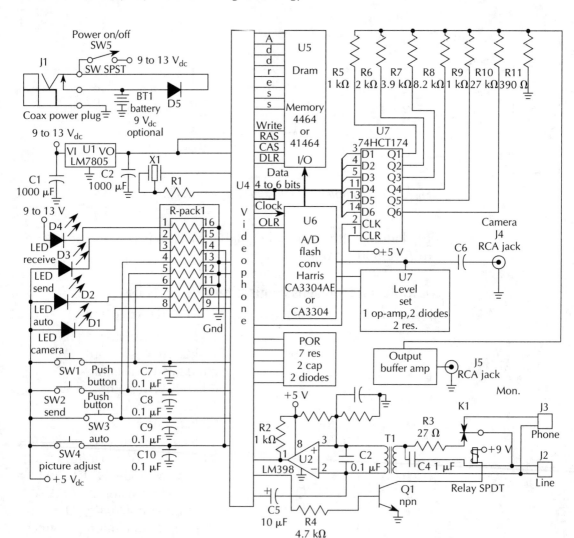

**Fig. 3-23** *Videophone block diagram.*

from PMC Electronics. This wonder of integration combines the equivalent of over 150 chips into one picturephone chip.

A picturephone system consists of two videophone stations connected via radio or a twisted wire pair. The video source for the picturephone chip consists of a fast-scan composite-video camera or a VCR. The display for the videophone can be a composite computer monitor or a TV set with an rf modulator. The resolution of a picturephone is over 50,000 pixels with up to 50 gray-scale levels.

The vidram system has a much greater resolution than any other videophone available. The videophone station consists of a single double-sided circuit board with the videophone chip, an A/D flash converter chip, a memory chip, output buffer op amp, regulator, speed circuitry, resistors, capacitors, matching transformers, switches, and chassis box. A camera and monitor are not included. The videophone system captures the odd field of a two-field interlaced frame and displays, and sends a 200 × 242 ½-line plus blanking and flyback information to the dynamic RAM memory, which can be a 64K chip or a 256K RAM. A 1-MB memory could also be used to store numerous pictures in its memory. The proprietary pulse-width modulation scheme sends data rates of 32 kb/s. Each baud is one pixel and conveys 6 bits of gray scale per baud. Since 1 baud is ½ cycle, 12-bits per cycle are conveyed almost instantaneously.

The videophone has four modes of operation. First, the capture mode takes the picture from your VCR or camera and stores it in the memory, and displays the picture on your monitor. The second mode extracts the video image from memory and sends the picture down the phone line at speeds from 8 to 36 seconds per picture, as determined by the user. The receive mode captures a picture from the distant station and automatically displays the incoming picture. The fourth mode is auto-sending, and captures still images and then sends them every 40 seconds. This unique feature allows for some interesting video surveillance systems. I will present them later. The videophone can be built from a PMC kit, or PMC also offers a completed version. The videophone can be constructed for under $150, less monitor and camera. PMC offers some new improvements that include color video and printing of received pictures. Video with conversation will be offered in future versions.

## Gas-sensing technology

Gas-sensing technology has progressed with rapid advancement in the last few years. New gas-sensing technologies detect oxygen, combustible gases, organic vapors, CFCs, alcohols, and toxic gases. New sensing techniques can be applied to ventilation control, fire alarms, leak detectors, auto emission controls, automatic cooking controls, fermentation controls, and gas analyzers. Under development are $NO^x$, ozone and thin-film multielement detectors that can differentiate between various blends of coffee, tobacco, and alcohols.

The mainstay of solid-state gas sensing has been the sintered tin dioxide ($SNO^2$) sensors, which detect gases through an increase in electrical conductivity when reducing gases are absorbed on the sensor's surface. Table 3-1 illustrates a wide selection of solid-state gas sensors. These sensors have excellent stability and performance, which allows them to be used in numerous applications that require long life, reliability, resistance to shock, and low cost. The earliest solid-state detectors were used to detect combustible gases. However, today's sensors can detect all types of gases and odors, as you will see. Various gas sensors are shown in Fig. 3-24.

Table 3-1  Gas sensors.

| Category | Model | Typical Detection Ranges | Features |
|---|---|---|---|
| Combustible Gases | TGS109 | LP-Gas (propane, butane) 500~10000 ppm | • 100-V circuit voltage type<br>• Large output signal<br>• Precalibrated sensor modules are available (see Table 3-2) |
| | TGS109T | Natural gas 500~10000 ppm | |
| | TGS813 | General combustible gases | • For various combustible gas detection |
| | TGS816 | 500~10000 ppm | • TSG816: heat-resistant type |
| | TGS842 | Methane, propane, butane 500~10000 ppm | • Low sensitivity to noise gases |
| | TGS815 | Methane 500~10000 ppm | • Small dependency on humidity and temperature |
| | TGS821 | Hydrogen 50~1000 ppm | • High selectivity and sensitivity to hydrogen |
| Toxic Gases | TGS203 | Carbon monoxide 50~1000 ppm | • High sensitivity and selectivity to carbon monoxide<br>• Microprocessor and hybrid IC are available (see Table 3-3) |
| | TGS824 | Ammonia 30~300 ppm | • High sensitivity to ammonia |
| | TGS825 | Hydrogen sulphide 5~100 ppm | • High sensitivity to hydrogen sulphide |
| Organic Solvents Detection | TGS822 TGS823 | Alcohol, toluene xylene, etc. 50~5000 ppm | • High sensitivity to alcohol and organic solvents TSG823: heat-resistant type |
| CFCs (Chlorofluorocarbons) | TGS830 | R-113, R-22 100~3000 ppm | • High sensitivity to various CFCs<br>• Good cross sensitivity |
| | TGS831 | R-21, R-22 | • Quick response to R-21, R-22 |

**Table 3-1 Continued.**

| | | | |
|---|---|---|---|
| | TGS831 | R-21, R-22<br>100~3000 ppm | • Quick response to R-21, R-22 |
| Odor Detection | TGS501 | Sulphur compounds<br>0.1~10 ppm | • Ultrahigh sensitivity to sulphur compounds<br>• Low power consumption<br>• For intermittent detection |
| Ventilation Controls | TGS100<br>TGS800 | Air contaminants<br>(cigarette smoke,<br>gasoline vapors, etc.)<br>Less than 10 ppm | • Highly sensitive detection of contaminants in air with microprocessors (see Table 3-3)<br>• Precalibrated sensor modules are available (see Table 3-2) |
| Cooking controls | TGS880 | Volatile gases, vapors from food (gas, humidity, smoke and odor) | • Total gas detection in cooking process |
| | TGS881 | | • TSG881: heat-resistant ceramic housing |
| | TGS883T | Humidity, volatile gas and vapors from food | • High sensitivity to humidity |

**Fig. 3-24** *Typical gas sensors.*

The operation of the tin-dioxide sensors begins with heating the detector to a high temperature via an internal heater. Free electrons flow easily through the grain boundaries of the tin dioxide in the absence of oxygen. In clear air, oxygen traps free electrons by its electron affinity. This oxygen is absorbed by the tin-dioxide particles, forming a potential barrier in the grain boundaries. This potential barrier in the grain boundaries (EVs in air) restricts the flow of electrons, causing the electrical resistance to increase. When the sensor is exposed to an atmosphere containing reducing gases such as CO or methane, the tin-dioxide surface absorbs these gas molecules and causes an oxidation process. This lowers the potential barrier, allowing electrons to flow more easily, thus reducing the electrical resistance of the sensor. Figure 3-25 illustrates this process of absorption on the sensor's surface. The reaction between gases and the surface oxygen differs, depending primarily on the sensing elements' temperature and the activity of the sensor's material composition. Various sensor designs are produced through different cross sensitivities and by selecting suitable combinations of sensor materials and element temperatures, as mentioned.

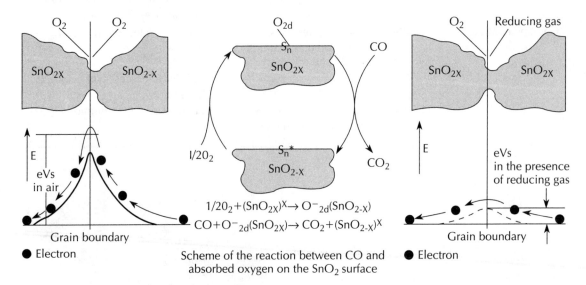

**Fig. 3-25** *Gas absorption on sensor surface.*

The series 1 sensors are designed to detect combustible gases such as propane, butane, or natural gas, with sensitivities as low as 500 ppm. These sensors have been used primarily for domestic gas-leak detectors as well as in recreational vehicles. These sensors detect natural gas or methane, and a second model sensor detects propane, butane, or LPG gases. These sensors are available in precalibrated or uncalibrated types. The series 1 sensors have two coiled electrodes made of an iridium/palladium alloy, which are encapsulated inside of the sintered $SNO^2$ sensor element. One or both of these electrodes are used as the heater assembly, as shown in Fig. 3-26. These sensors are usually operated from 100 V with heater voltages of 1 V, as shown in the wiring diagram of Fig. 3-27.

Gas-sensing technology 109

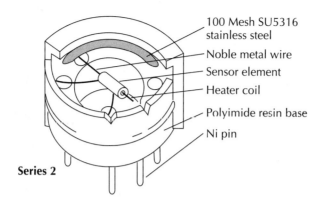

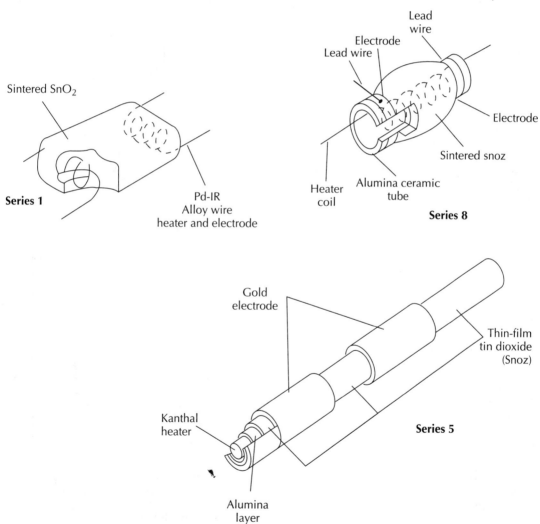

**Fig. 3-26** *Diagram of typical gas sensors.*

**1 series**

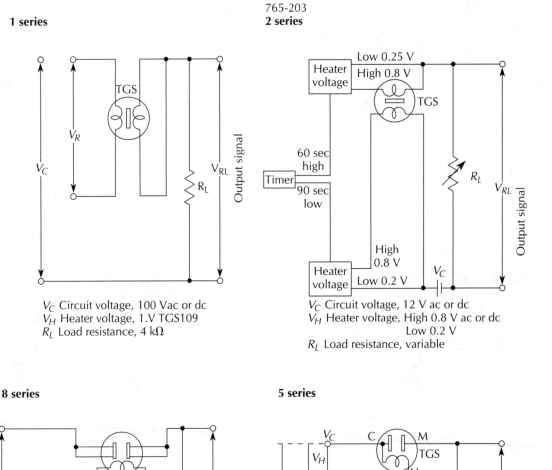

$V_C$ Circuit voltage, 100 Vac or dc
$V_H$ Heater voltage, 1.V TGS109
$R_L$ Load resistance, 4 kΩ

**765-203
2 series**

$V_C$ Circuit voltage, 12 V ac or dc
$V_H$ Heater voltage, High 0.8 V ac or dc
                         Low 0.2 V
$R_L$ Load resistance, variable

**8 series**

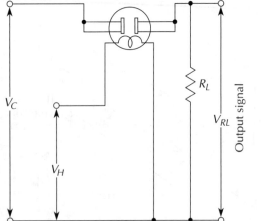

$V_C$ Circuit voltage, 24 V max ac or dc
$V_H$ Heater voltage, 5 V ac or dc
$R_L$ Load resistance, variable

**5 series**

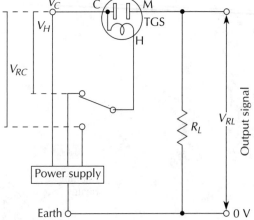

$V_C$ Circuit voltage, 5 V max ac or dc
$V_H$ Heater voltage, 0.55 V ac or dc
$V_{HC}$ Heat cleaning voltage, 0.67 V
$T_{HC}$ Heat cleaning time, 10 sec to 5 min

**Fig. 3-27** Basic gas-sensor measuring circuits.

The series 2 type sensors, such as the TGS-203, are designed for carbon-monoxide detection. These highly sensitive sensors can detect a 50-ppm concentration of carbon monoxide. A built-in temperature-compensation circuit minimizes the sensitivity fluctuations caused by ambient temperature. The sensing element is an n-type metal-oxide semiconductor, in which the electrical conductivity increases when in contact with deoxidizing gases. The sensitivity depends upon the temperature at which the element is heated. For CO detection, the best temperature is below 100°C, which is lower than that of other gases such as butane or methane. The CO sensors can be readily influenced by water vapor in the air. Therefore, the sensing element is heated alternately from a high to low voltage. During the high-temperature period, water vapor and other gases are removed from the sensor's surface. During the low-voltage period, the sensor is allowed to detect CO gases. The heater voltage cycles between 0.80 V for 60 seconds and then to a low voltage of 0.25 V for a 90-second period. The nominal circuit voltage for the TGS203 sensor is 12 V, as shown in Fig. 3-27. The 200 series sensors are placed inside of a 17-mm diameter can, as seen in Fig. 3-4. Both upper and lower openings of the sensor case are covered with a flameproof double layer of 100-mesh stainless-steel gauze. The mesh prevents any type of spark from igniting an explosive gas mixture.

Sensitivities of the gas detectors discussed are shown in Figs. 3-28 and 3-29. In these figures, the x axis is normalized by the sensor resistance $R^o$ at a specified condition for each model. The y axis is indicated as the sensor's resistance ratio Rs/Ro, instead of the sensor's resistance value $R^s$.

A custom integrated circuit with the proper voltages and heating cycle Rs for the model TGS203 sensors is shown in the block diagram Fig. 3-30. The block diagram of the FIC5401 custom IC from Figaro Engineering illustrates a 4-bit microprocessor with built-in constant-current supply for the heaters. The output signal from the sensor is processed by the internal comparator, which uses a precise threshold adjustment control. When the sensor signal exceeds the threshold, an alarm output is triggered. A trouble-signal output is also provided to indicate any type of disconnect faults or open heater circuit conditions. The built-in temperature-compensation circuit decreases the effect of the ambient temperature fluctuations. The custom IC unit operates from a 5-V source, which allows for operation from a 6-V battery.

The custom IC module provides two alarm outputs. Alarm A can directly drive a piezo buzzer or an LED from pin 40. The sec-

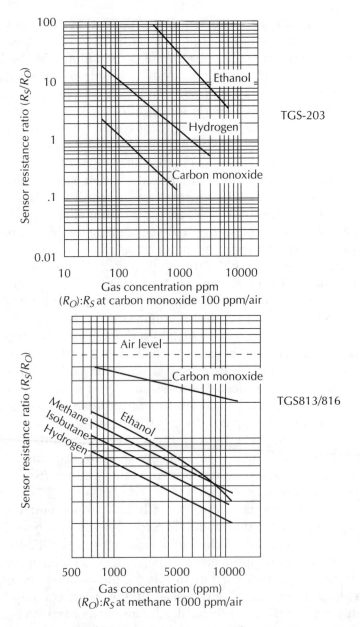

**Fig. 3-28**  *TGS203/813 sensitivity characteristics.*

ond alarm output, B, can trigger an alarm control panel. A complete CO detection system can be created by using the custom FIC5401 module, shown in Fig. 3-31. A temperature-compensating thermistor is connected to pin 17 and $V_{CC}$. The heater voltages for the sensor are applied to pins 21, 23, 24, and 25. Power to the IC module is found on pins 8, 11, 19, 25, 30, and 34, while

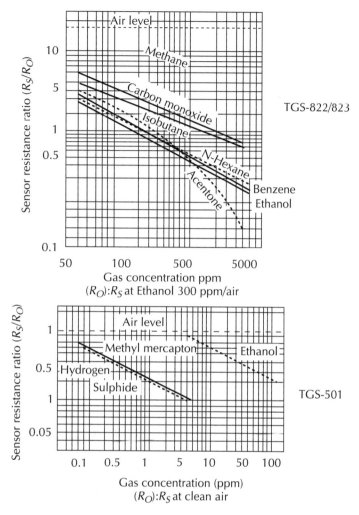

**Fig. 3-29** *TGS822/501 sensitivity characteristics.*

ground pins can be found on pins 5, 14, 20, 25, and 31. The IC module is ideal for constructing a complete CO detection system.

Figaro's series 5 model TGS550 sensor is a low-current, thin-film $SNO_2$ sensor developed to detect offensive mouth odors, particularly methyl mercaptan. The sensor has three main components (see Fig. 3-26). The Kanthal heating wire is covered with an alumina insulation layer, a thin layer of $SNO_2$, and a pair of 90 LD electrodes, which form the active elements of this sensor. The series 5 detectors operate from a 5-V source with a 0.55-V heater voltage, as seen in the wiring diagram of Fig. 3-27. This sensor can also detect hydrogen sulfide down to 0.1 ppm.

*114  New Sensors, ICs, and Gas-Sensing Technology*

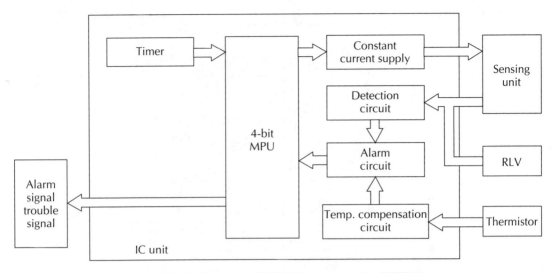

**Fig. 3-30** Block diagram of FIC5401 custom IC—TGS203 sensor.

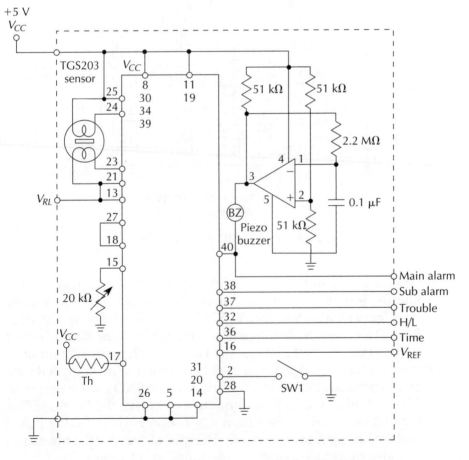

**Fig. 3-31** Custom FIC5401 IC electronics for TGS203 sensor.

The series 8 gas sensors, such as the TGS813, primarily detects methane, propane, butane, and LPG gases. These sensors are used as domestic gas-leak detectors and as gas-leak detectors in recreational vehicles and marine applications. This series has very good sensitivity, with little interference from noise gases, which can cause false alarms. The 813 series sensors are low cost and highly reliable. They are composed of a heater coil located inside of a ceramic former. The heater coil is a 60-micron wire with a 30-$\Omega$ resistance. The sintered $SNO_2$ semiconductor material is then deposited on the ceramic former assembly. The TGS813 sensors are assembled in the same 17-mm can as is the 200 series sensors. The 800 series detectors are affected by changes in atmospheric temperature and humidity. These effects cannot be entirely eliminated, but they can be significantly reduced by using thermistor compensation. Because of detectors' sensitivity characteristics, the action of the sensor will vary according to the type and concentration of the gas it is detecting. The proper alarm set point for a particular sensor should, therefore, be determined after considering the following characteristics

- Where is the sensor to be located or installed
- Purpose of the detector, gas-leak, fan-control, or air-monitoring system
- Operation of the detector to control fan, lights, or sound
- Type of gas detection used for.

A complete methane-gas detection system is shown in Fig. 3-32. This circuit was designed to be used as a domestic leak detector. The detector circuit will sound an audible alarm when a gas concentration exceeds the threshold setting. A regulator provides a constant-current supply for the sensor's heater and the detection circuit. The voltage $V_{RL}$ across resistor R1 and R8 enters the comparator IC via the noninverting input of the IC, from the sensor. Once the detector signal exceeds the threshold setting, an audible alarm sounds. Since ambient temperature and humidity can affect the alarming point, as illustrated in Table 3-2, a negative-temperature-coefficient thermistor is used to stabilize this variation upon the sensor. The chart in Table 3-1 depicts a constant humidity with a varying temperature. The results are shown for both a noncompensated as well as a compensated sensor. The temperature-compensation circuit consists of R2, R3, and R9.

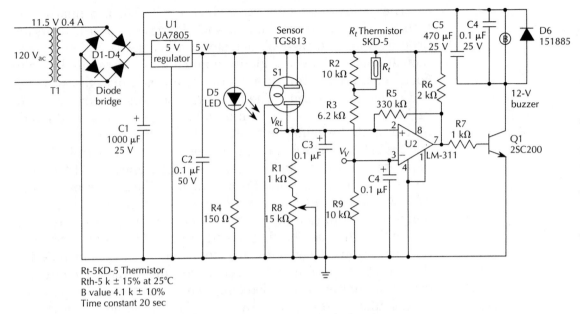

**Fig. 3-32** Methane-gas leak detector.

Rt-5KD-5 Thermistor
Rth-5 k ± 15% at 25°C
B value 4.1 k ± 10%
Time constant 20 sec

Table 3-2  Setpoint variations—TGS813 sensor.

| Temperature °C | Humidity %RH | Circuit with temperature compensation | Circuit without temperature compensation |
|---|---|---|---|
| 20 | 65 | 3000 | 3000 |
| −15 | 65 | 2950 | 6000 |
| 5 | 65 | 3100 | 4800 |
| 35 | 65 | 2600 | 1400 |

Calibrating the detection circuit is accomplished by adjusting the R8 potentiometer at 20°C temperature, with a relative humidity of 65%. The $V_{RL}$ voltage should read 2.5 V at TP1, in a 3000-ppm concentration of methane. When installing the TGS813 sensor, two caveats should be observed. Never place the sensor in a drafty or windy location, and never place the sensor near a transformer, including transformer T1. The TGS813 series detectors are available in two classifications—methane or propane. The methane detector is used for natural gas and the propane model is used for propane, butane, and LPG gases. A small colored dot on the sensor's body denotes the particular type.

The TGS822/823 gas sensors are highly sensitive to organic-solvent vapors from alcohol, toluene, and xylene (see sensitivity in Fig. 3-29). These sensors are particularly sensitive to ethanol vapors down to 50 ppm. The TGS822 series sensors are used for alcohol and toluene detection, and the TGS823 series are used for xylene detection. These sensors are also sensitive to temperature and humidity variations, and a compensation thermistor should be incorporated into the detection system design. The TGS822/823 sensors are packaged in the same 17-mm can as in the other 800 series sensors. The maximum circuit voltage for the TGS822/823 sensors is 24 V, with a maximum heater voltage of 5 V, as illustrated in the wiring diagram in Fig. 3-27.

After long storage periods or period of inactivity, both the 200 and 800 series sensors require a 2-minute stabilization period before accurate gas detection can be attempted.

The circuit in Fig. 3-33 illustrates an alcohol breath analyzer using the TGS822 series sensor. The breath analyzer features four LEDs to indicate alcohol concentrations in a person's breath. The

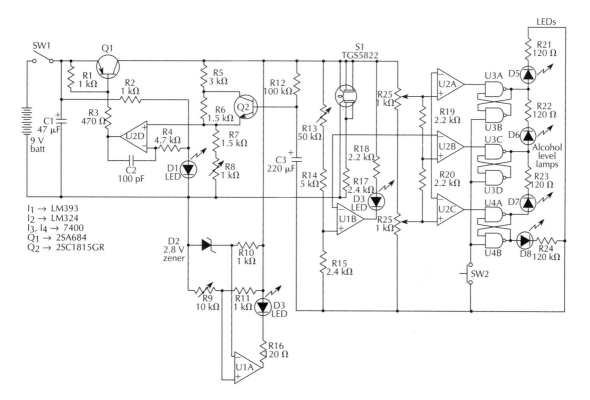

*Fig. 3-33* Alcohol breath analyzer.

circuit incorporates two 1000-Ω trim pots for calibrating the display. The breath analyzer can be powered by a standard 9-V battery. The circuit uses a regulator to provide a constant source current to the sensor's heater, as well as a low-battery-voltage indicator. The circuit also features a heat-cleaning circuit for the sensor, as well as a ready lamp that indicates the time of optimum sensitivity when the detector is ready to "sniff" for alcohol.

The GS oxygen sensor, shown in Fig. 3-34, is a newly developed galvanic-cell-type detector. These new oxygen sensors have already found their way into the medical field for anesthetic instruments, spirophores, oxygen incubators, and numerous applications in the food industry. These new sensors have also been used in air purifiers, air conditioners, and refrigerators. The GS series oxygen sensors operate by allowing oxygen molecules to diffuse through a non-porous teflon membrane into the electrochemical cell. The gas is then reduced at the gold electrode in the cell assembly. The current flowing between the lead anode and the gold cathode is proportional to the oxygen concentration, in a gas mixture. The changes in voltage output is represented di-

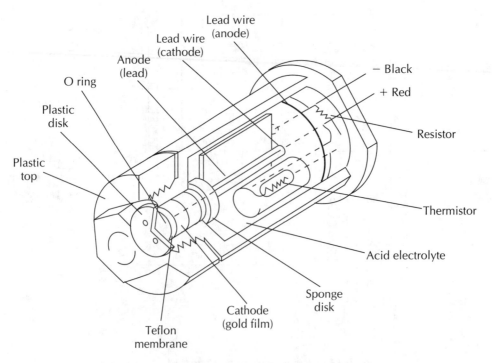

*Fig. 3-34* Structure of KE-25/KE-50 oxygen sensor.

rectly by the oxygen concentration which can vary from 0 to 100%. The GS sensors are accurate to 1% with excellent reproducibility. These sensors are available in two relatively low-cost models. The model KE-50 has a 10-year life span.

Detection of low concentrations of air pollution, e.g., cigarette smoke and cooking fumes, are possible with a combination of the TGS100/TGS800 sensors coupled with the 5603/6604 custom microprocessors. The microprocessor calculates the average value of the sensor's resistance in ambient air over a certain period of time, and it renews the base level. This avoids any influence from humidity, temperature, or environmental changes that may have taken place. This method is quite effective for automatic cooking controls, ventilation, and air purifiers. Air-evaluation modules are available to help the engineer design new automatic controls. The AM100 module is available for use with the TGS100 sensor, and the AM800 is available for the TGS800 sensor. As noted earlier, during the process of cooking, various volatile materials are emitted from food, i.e., humidity, alcohol vapor, and combustion gases. The sensor's resistance changes are due to the presence of those gases. The emission of volatile materials differs, depending on the type of food and the cooking methods (see Fig. 3-35). The cooking time can be automatically controlled by using the gas sensor's output signals. Other gas sensors available from Figaro include the TGS830 series detectors, which detect CFCs in 100-ppm concentrations, and the TGS821 series, which can sense ammonia down to 30 ppm.

Teknetron, another leader in thin-film gas sensors, is developing a number of new thin-film metal-oxide sensors. Initial prototypes have shown great promise for a new generation of low-cost sensors. Teknetron has recently developed a back cell-type-microelectrochemical gas sensor, shown in Fig. 3-36. It consists of a substrate material with an opening across which a gas-permeable sensing electrode is placed. The front side of the electrode is coated with an electrolyte. The gas diffuses from the back side into the sensor. The design is unique in that the material to be detected can diffuse directly to the electrode without having to bridge a liquid path, as in most electrochemical sensors of the past. Thus, a fast linear response is produced, which is an order of magnitude faster than in the sensors currently available. Electrochemical sensors, in general, have lower power requirements and higher selectivities than their semiconductor cousins. Howver, the classical drawbacks of electrochemical sensors have been slow response times and short life.

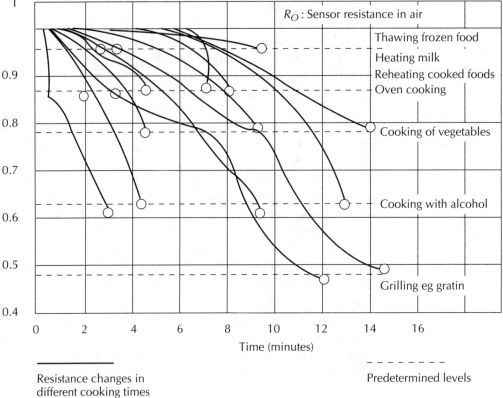

**Fig. 3-35** *Variations in sensor resistance ratios in microwave-oven cooking.*

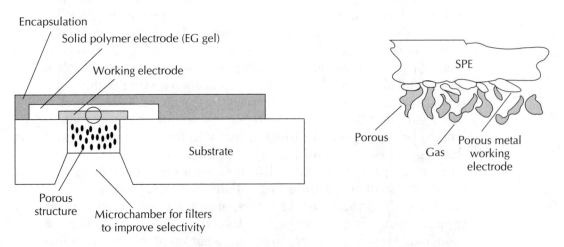

**Fig. 3-36** *Back-cell microelectrochemical gas sensor.*

Experimentation with titanium dioxide, zinc oxide, iron oxide, and, more recently, bismuth molybdate have shown good promise for future semiconductor-type sensors.

Investigators at the Edison Technology Center have demonstrated that thin sensors can be used to detect ethane, halothane, pentane, and ethyl ether, all of which are used in hospital operating rooms.

Dr. Chang of the General Motors Research Laboratories has investigated various types of solid-state gas sensors. He has determined that certain additives to the $SNO_2$ thin films have strong effects upon their overall characteristics. A pure $SNO_2$ film was found to show a high sensitivity/selectivity to $NO_x$ in air, but a thin overlay of palladium film and gold over the $SNO_2$ layer slightly depresses the sensitivity of $NO_x$, but greatly enhances the sensitivity to propylene.

A new CHEMFET (chemical field-effect transistor) consists of a chemically sensitive layer placed onto the gate of an FET. Palladium has been found to be sensitive to hydrogen and hydrocarbon gases, and it is typically coated onto the gate of an FET. Altered FET gate structures have been successfully used to detect CO and methane gases. CHEMFETs can also be used to sense multiple species of gases, and they can be fabricated quite inexpensively. The experiments with CHEMFETs have proved very fruitful, and production is expected within two years.

New ISFETs (ion-sensitive field-effect transistors) have been developed recently, and they look very promising as well. These detectors use fatty acids that form monolayer surfaces. These sensors are intended to be used as ion-selective membranes on the ISFET insulators. Stearic-acid layers have also been used on the quartz substrates of new surface acoustic-wave sensors now being experimentally used in miniature gas chromatographs. The new ISFETs appear very interesting and promising for future sensing applications.

Future trends in gas-sensing technology include thin-film $SNO_2$ multisensor arrays. Experimental chemical-vapor-sensing devices have been algorithmically trained to distinguish up to 12 slightly different varieties of tobacco smoke, coffee aromas, and alcohol blends. The latest research also includes nitrogen-dioxide sensors, which can detect these gases down to 20 parts per billion. New ozone detectors are also on the horizon. Traditionally, ozone detectors have been complicated and very expensive. Dual sensors with special wavelength filters are amplified and sent to a microprocessor that computes a number of calculations before the results can be displayed. The new ozone detectors will greatly reduce the size and cost of these detection systems. The new sensing technologies are very exciting and we will most likely benefit from the current research in the near future.

## Methane gas leak detector parts list

| Quantity | Part | Description |
|---|---|---|
| 2 | R1, R7 | 1-kΩ, ¼-W resistor |
| 2 | R2, R9 | 10-kΩ, ¼-W resistor |
| 1 | R3 | 6.2-kΩ, ¼-W resistor |
| 1 | R4 | 150-Ω, ¼-W resistor |
| 1 | R5 | 330-kΩ, ¼-W resistor |
| 1 | R6 | 2-kΩ, ¼-W resistor |
| 1 | R8 | 15-kΩ potentiometer |
| 1 | Rt | Thermistor, 5 kΩ ± 15% at 25°C |
| 1 | C1 | 1000-µF, 50-V electrolytic capacitor |
| 4 | C2, C3, C4, C6 | 0.1-µF, 50-V capacitor |
| 4 | D1-D4 | 1N4002 silicon diode |
| 1 | D5 | LED |
| 1 | D6 | 1S1885 diode |
| 1 | Q1 | 2SC200 transistor |
| 1 | U1 | UA7805 regulator |
| 1 | U2 | LM311 op amp |
| 1 | B | Piezo buzzer |
| 1 | T1 | Transformer 110V–11.5 $V_o$, 4A |

## Alcohol breath analyzer parts list

| Quantity | Part | Description |
|---|---|---|
| 5 | R1, R2, R10, R11, R25 | 1-kΩ, ¼-W resistor |
| 1 | R3 | 470-Ω, ¼-W resistor |
| 1 | R4 | 4.7-kΩ, ¼-W resistor |
| 1 | R5 | 3-kΩ, ¼-W resistor |
| 2 | R6, R7 | 1.5-kΩ, ¼-W resistor |
| 2 | R8, R25 | 1-kΩ, trim pot |
| 1 | R9 | 10-kΩ, trim pot |
| 1 | R12 | 100-kΩ, ¼-W resistor |
| 1 | R14 | 5-kΩ, ¼-W resistor |
| 2 | R15, R17 | 2.4-kΩ, ¼-W resistor |
| 6 | R16, R18, R21, R22, R23, R24 | 120-Ω, ¼-W resistor |
| 2 | R19, R20 | 2.2-kΩ, ¼-W resistor |

| | | |
|---|---|---|
| 1 | C1 | 47-µF, 25-V electrolytic capacitor |
| 1 | C2 | 100-pF, 25-V capacitor (disk) |
| 1 | C3 | 220-µF, 25-V electrolytic capacitor |
| 1 | D1, D3-D7 | Red LEDs |
| 1 | D2 | 2.8-V zener diode (NTE5064A) |
| 1 | Q1 | 2SA684 transistor |
| 1 | Q2 | 2SC1815GR transistor |
| 1 | U1 | LM393 op amp |
| 1 | U2 | LM324 op amp |
| 2 | U3, U4 | 7400 NAND gate |
| 1 | SW-1 | SPST switch |
| 1 | BATT | 9-V battery |

# ❖ 4
# Computer interfacing

THE PERSONAL COMPUTER HAS CHANGED MOST OF OUR LIVES more than any other invention in our recent history. Computers are capable of many applications in addition to word processing and games. Computers can also be used to sense and control the world around us. The joystick port and serial port both provide an inexpensive means to collect remote data from resistive sensors, switches, and voltages, as well as to control objects remotely. Recently introduced, low-cost, multichannel, analog-to-digital converter cards allow almost everyone to construct data acquisitions and control systems.

## Joystick interfaces

There are numerous approaches to interfacing and collecting data from analog sensors. I will discuss some low-cost and often overlooked methods and ideas for interfacing sensors to personal computers. One simple, interesting method utilizes the serial game-port card. Potentiometer joysticks have two potentiometers that provide proportional outputs with respect to the position of the joystick. The two gimbel-mounted potentiometers form an x-y coordinate axis system. You can easily interface resistive-type sensors or voltages in place of the joysticks. Photoresistors can be used to measure light levels, and thermistors can measure temperature levels. Resistive sensors can also be used to detect position, tilt angles, liquid levels, and directional measurements.

Figure 4-1 illustrates two common interface connections to game-port cards for an IBM Personal Computer as well as other clone computers. Any voltage from 0 to 5 $V_{dc}$ or any resistive sensor can be connected in place of the joysticks. Most joystick potentiometers use a 100-k$\Omega$ linear-type potentiometer.

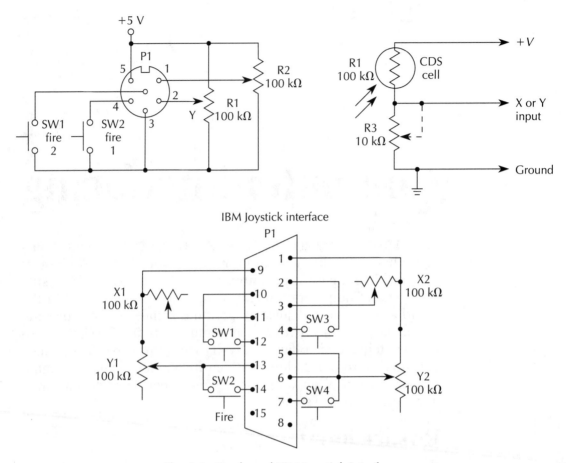

**Fig. 4-1** Tandy and IBM joystick interfaces.

You can easily measure light levels by obtaining a cadmium-sulfide light cell from Radio Shack (RS 276-1657). By substituting a thermistor, you can measure temperature variations (RS 271-110). By substituting a long-shaft 100-kΩ potentiometer, you can construct a liquid-level detector. First, drill a hole through the potentiometer shaft, place a small, threaded shaft through the hole in the potentiometer, and place a nut over the threaded shaft, attaching the shaft to the threaded rod. At the opposite end of an 8-inch threaded rod, you can attach a small, lightweight ball, to act as a float. As the liquid level changes, the ball moves up and down, thus changing the resistance values. A simple wind-direction indicator can be fabricated by using the same setup, but instead of a ball attached to the threaded rod, a plastic tail is attached to act as a wind vane. For this sensor, remove the potentiometer end stops so that the pot can move through 360 degrees. The only drawback to this arrangement is the electrical dead zone of the potentiometer.

You can also interface many different types of switches to the four digital inputs of the game-card input. You can measure the speed of many different objects by connecting a magnetic reed switch to one of the digital switch inputs. Speeds up to 300 rpm can be monitored easily. Mercury tilt switches, leaf switches, vibration switches, and level switches can be connected to the switch inputs on the game card.

Game-card inputs can be read by assembly-language programs, but a simpler method to read the joysticks can be implemented by using the STICK or STRIG commands in BASIC. The STRIG command is used to read the digital-switch inputs, and the STICK command is used to read the analog potentiometer or joysticks. The ON STRIG command can be used to implement a digital interrupt.

The IBM game-port adaptor provides a method of resistance conversion that changes resistance to pulse duration. Resistance of each joystick varies the duration of each of the timers, such as a 555/556 chip. The one-shot timers are fired by an output command to port address 0201 hex.

An IN command reads the values from the timers, as well as the status of the fire buttons. The time interval for each of the one-shots (i.e., the timeout period) is proportioned to the joystick's actual resistance. The IBM computer contains the clocks and registers that convert duration to digital numbers. A more thorough discussion of the game port can be found in the IBM *Tech Reference* manual.

Another approach to interfacing sensors to computers is shown in Fig. 4-2. This circuit essentially involves constructing your own joystick interface card to obtain the results just discussed. Two joysticks are coupled to a dual 556 IC timer. The timers are fed to a 74L5244 and then to a pair of 74LS32 gates. These gates are then connected directly to the IBM computer bus. This card allows any voltages from 0 to 5 $V_{dc}$ or resistive sensors.

## Serial interface

Yet another method to interface sensors to the PC allows signals to be coupled directly to an RS-232 asynchronous input card instead of a game-port card. Figure 4-3 illustrates this simple method of interfacing. Resistive sensors are fed to a timer chip, and timing durations are once again converted to digital pulses.

*128* Computer Interfacing

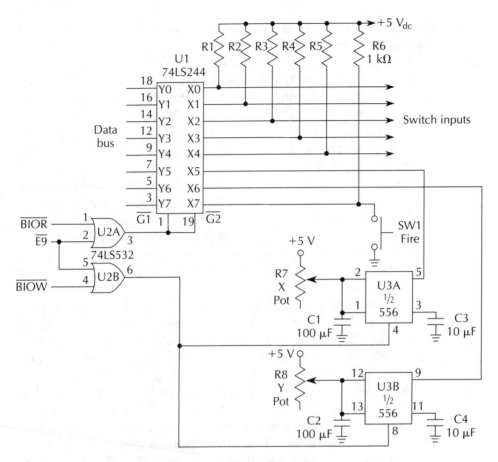

**Fig. 4-2** Bus joystick interface card.

The timer outputs are connected to a JFET op amp, which inputs values of time duration directly to an RS-232 input connector. Further details, as well as a software routine, are covered in an interesting article in the November 1990 *PROBE* magazine entitled, "Home Computer WX Station".

## Mouse and trackball interface

The trackball is a graphic-input control device that can also input analog data. The trackball consists of a ball mounted in a base with part of the ball's surface exposed. Movement of the ball causes two optical encoders to generate pulses that indicate both speed and direction of rotation. Most trackballs produce between 50 and 550 pulses per revolution. The ball lies on two perpendicular rods for both the x and y axis. An optical interrupter forms a pulse train by breaking the path of light between a micro

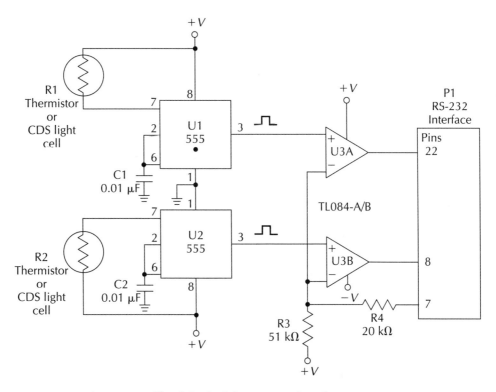

**Fig. 4-3** *Serial computer interface.*

LED and a photodiode. The mouse is a similar device with an upside-down trackball. The mouse and trackball are connected to a computer in two ways. In one method, the mouse connects to the asynchronous serial port. The other method of interface is the parallel bus type, which uses a special interface card plugged into the PC bus.

A typical trackball is shown in Fig. 4-4. The mouse or trackball can provide an inexpensive method of measuring speed and direction. The mouse or trackball can be disassembled, and the optical disk rods can be extended via a shaft coupler or extension rods, to allow you to interface with the outside world. This simple approach can save you a considerable amount of money, since low-cost mouse units are available for less than $30. The mouse or trackball is a clever way to solve counting and direction measurements.

## Analog-to-digital interface

The last method of interfacing analog sensors to an IBM clone is to use a true analog-to-digital conversion card. Figure 4-5 illus-

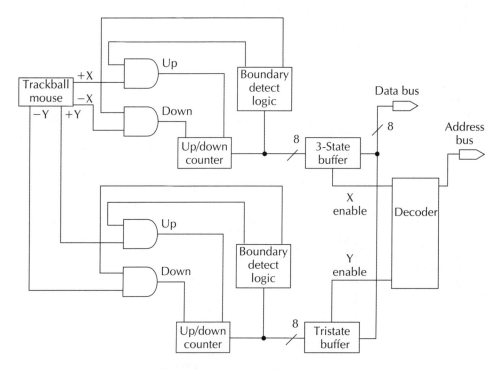

**Fig. 4-4** Mouse/trackball interface.

trates an 8-bit A/D circuit that can be built for less than $40. Resistive sensors or analog voltages can be applied to this card. The A/D circuit can be constructed on a prototyping card inserted inside your computer. If you choose not to construct your own A/D card, you might consider purchasing a low-cost A/D card. The 8- to 10-bit A/D cards have become quite reasonable in the past few years and can be purchased for $85 or less. The computer has become very important in collecting and processing data, and has replaced chart recorders in many applications. Precision measurements can be taken with considerable ease.

## Tandy joystick port parts list

| Quantity | Part | Description |
|---|---|---|
| 1 | R1 | Resistive sensor, 50–100 kΩ |
| 1 | R2 | Resistive sensor, 50–100 kΩ |
| 1 | R3 | 10-kΩ potentiometer |
| 1 | SW1, SW2 | NO pushbutton or on-off sensor switches |
| 1 | P1 | Joystick connector |

Analog-to-digital interface   131

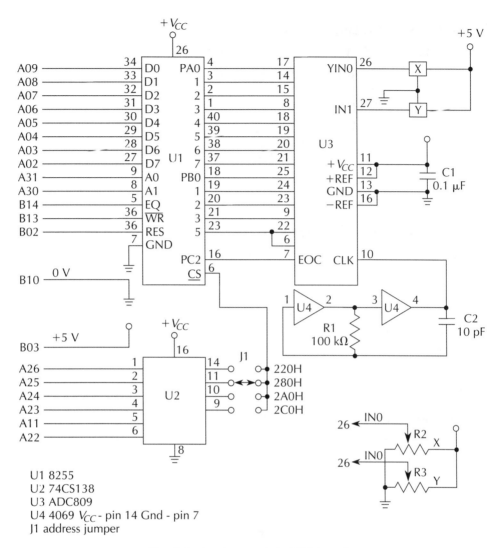

**Fig. 4-5** Analog-to-digital PC interface.

## IBM joystick port parts list

| Quantity | Part | Description |
|---|---|---|
| 2 | X1, Y1 | 50–100-kΩ resistive sensors |
| 2 | X2, Y2 | 50–100-kΩ resistive sensors |
| 2 | SW1, SW2 | NO switch sensors (reed switch) |
| 2 | SW3, SW4 | NO switch sensors (reed switch) |

## Bus joystick interface parts list

| Quantity | Part | Description |
|---|---|---|
| 6 | R1–R6 | 1-k$\Omega$ ¼-W resistors |
| 2 | R7, R8 | 100-k$\Omega$ resistive sensors |
| 2 | C1, C2 | 100-nF capacitors (disk) |
| 2 | C3, C4 | 10-nF capacitors (disk) |
| 1 | U1 | 74LS244 IC |
| 1 | U2 | 74LS32 IC |
| 1 | U3 | 556 dual IC timer |
| 1 | SW1 | NO switch contacts (reed switch) |

## Serial computer interface parts list

| Quantity | Part | Description |
|---|---|---|
| 2 | R1, R2 | 50–100-k$\Omega$ resistive sensors |
| 1 | R3 | 51-k$\Omega$, ¼-W resistor |
| 1 | R4 | 20-k$\Omega$, ¼-W resistor |
| 2 | C1, C2 | 0.01-$\mu$F, 25-V capacitor (disk) |
| 1 | U1, U2 | 555 IC timer |
| 1 | U3 | TL084 op amp (Texas Instruments) |
| 1 | P1 | Serial-interface connector |

## Analog-to-digital PC interface parts list

| Quantity | Part | Description |
|---|---|---|
| 1 | R1 | 100-k$\Omega$, ¼-W resistor |
| 1 | R2 | Resistive sensor |
| 1 | R3 | Resistive sensor, 20–50k$\Omega$ |
| 1 | C1 | 0.1-$\mu$F, 25-V capacitor |
| 1 | C2 | 10-pF, 25-V capacitor (disk) |
| 1 | U1 | 8255 IC |
| 1 | U2 | 74LS138 IC |
| 1 | U3 | ADC809 IC |
| 1 | U4 | 4069 IC |
| 1 | J1 | Jumpers |

# Describing and surveying alarm systems

SENSORS ARE THE HEART OF ANY SECURITY SYSTEM. KNOWING when to use a particular sensor is very important, because each sensor has a specific intent and application. This chapter draws on the experience of alarm designers, installers, and the author to assist you in becoming aware of the advantages and pitfalls of the various sensors available. The first step in designing an efficient and successful security system is choosing the best sensor to suit each particular application.

Security-system designers and installers have selected switches and sensors that, over time, have proven themselves to be most reliable and cost effective. While most any sensor can be incorporated into an alarm system, particular sensors have become widely accepted and used. Presented in this section are the most commonly used sensors in the alarm community. I will discuss the advantages and disadvantages of these sensors.

## Window foil

One of the most widely used sensing devices in both commercial and industrial alarm systems is the ⅜-inch-wide metalized self-adhesive window foil (see Fig. 5-1). The window foil is carefully placed on the window pane. Then a coating of shellac or Varathane is applied over the foil to ensure that the foil remains affixed to the window glass. The ends of the foil are attached to self-adhesive terminator block. The window foil acts as a continuous loop circuit. Closed-loop sensing circuits offer the best protection because the circuit is always "supervised." A break in the loop circuit becomes immediately known when alarming the circuit. Window foil is an excellent deterrent, because it can be

**Fig. 5-1** *Window foil and terminal blocks.*

readily seen by any would-be intruder. The low-cost window-foil sensor is very effective, but it is very labor intensive to install. Window foil can be used in residential or home alarm systems, but many homeowners consider window foil very unattractive. As a consequence, other sensors are often used.

## Magnetic switches

The magnetic reed switch is probably the second most often used alarm sensor. It is available in many different styles and sizes (see Fig. 5-2). The magnetic reed switch consists of a magnet section and a glass-enclosed reed-switch unit. The two-piece reed switch is a low-cost, reliable sensor. This type of switch sensor never needs adjustment, and the contacts will never corrode. The switch is available in both normally closed and normally open outputs. The magnetic reed switch is often recessed into a doorjam so that it cannot be seen or tampered with by an intruder. The magnetic reed switch is widely used in alarm systems, but it, too, is quite labor intensive to install, especially if it is installed in a hardwire loop system. Many new alarm systems often incorporate magnetic reed switches inside a wireless transmitter, or provisions are provided to connect an external reed switch.

The wireless alarm system has gained enormous acceptance by alarm-system installers, because the wireless system is so much easier to install than conventional hardwire systems. Often a transmitter is placed at each door and window. Close-proximity doors and windows sometimes are connected to the same transmitter to save the cost of putting a transmitter at every window or door. Wireless transmitters are often connected to "space"

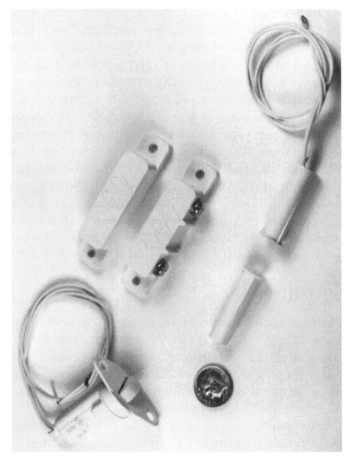

**Fig. 5-2**  *Magnetic switches.*

or interior protection devices such as motion detectors, which are generally used to protect the interior spaces within a building. Wireless transmitters are used in conjunction with a wireless receiver/control box. I will discuss the wireless system in more detail in the next section.

## Holdup switch

The panic or holdup switch had its roots in the banking industry. The first holdup switches were doorbell switches mounted under a teller's window. In the event of a holdup or confrontation, the teller would summon the bank manager or police department. The holdup switch then became widely used in retail stores, convenience stores, and gas stations—in fact, any place where cash is held. As crime moved into the suburbs, the panic switch became a regular part of residential alarm systems.

In hard-wired alarm systems, panic switches are strategically mounted throughout the home or office. The panic switch is often used by people living alone, by children left home alone, or by spouses who are home by themselves for long periods of time. In many instances, a panic switch is coupled to a portable transmitter that transmits to a compatible receiver unit. The receiver can be part of an alarm control box or a complete panic-alarm system can be separate from any other alarm system.

The wireless panic system is connected to an automatic telephone dialer. The receiver and dialer can be combined into one stand-alone panic alarm system, which automatically can call local relatives, parametics, or the police department. Panic or holdup switches have recently been installed in convenience stores to trigger cameras that take pictures of troublemakers or holdups, to identify the criminal to the police.

## Sensor mats

The *tape-switch mat* is relatively low cost and reliable, and is often used as a "space" or "spot" sensor for interior protection. The tape switch is normally a second line of defense after a perimeter alarm loop. The tape switch is usually placed under carpets to sense people walking through the home or office. "Spot" protection mats are 3- to 4-inch circular mats designed to protect specific valuables such as antiques, stereos, or TVs. Tape-switch mats are available in normally open or normally closed versions. They are often used as annunciators in retail stores to summon a clerk.

## Motion sensors

The *vibration detector*, in its simplest form, usually consists of a spring-steel lever arm with a small mass attached at one end. The lever arm acts as one of two contacts of a simple switch. The vibration sensor often is used instead of window foil to protect windows (see Fig. 5-3). The vibration sensor is often mounted on walls or epoxied to window glass. These simple, low-cost vibration sensors are available in both normally closed or normally open types and often used in inexpensive automotive alarm systems.

The major drawback to motion sensors is the tendency to use them at maximum sensitivity, which invariably causes them to false alarm. They are quite often used to protect window panes. Loosely fitted windows, storms or wind, or slamming doors can cause them to produce false outputs.

Motion sensors  *137*

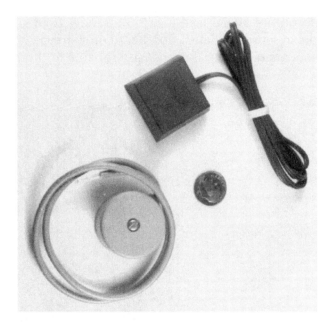

***Fig. 5-3*** *Window-vibration sensors.*

Vibration sensors are also available in more sophisticated forms, often using piezo-detecting methods as described in the project of the last chapter. A piezo elastomer is placed on a spring-steel arm and is free to move back and forth. The output of the detector is then directed to an amplifier and discriminator that triggers a relay (see Fig. 5-4). New audio window-breakage detectors use special discriminators that look for a specific number of pulses in a time window, before the sensor is allowed to produce

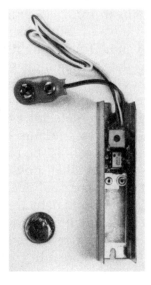

***Fig. 5-4*** *Piezo vibration sensor.*

an output signal (see Fig. 5-5). A new floor-strain sensor uses a piezo sensor to detect floorboard or staircase warpage. These sensors are relatively low cost and reliable for home installation.

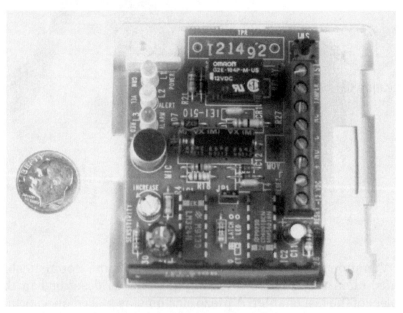

**Fig. 5-5** *Glass-breakage (sound) detector.*

*Interior "space" detection* sensors are more expensive than tape switches or foil. They are used to protect the inside areas of homes or buildings instead of perimeter type detectors. These sensors detect people once they have actually entered the premises.

The *light-beam sensor* has served the alarm industry for quite a long time and can be used for both perimeter or space protection. By mounting the light-beam detector on a long wall and facing the detector parallel to that wall, the detector can protect windows and doors along that perimeter. If the detector is pointed into the room area, it can provide "space" protection. The light-beam detector evolved from a simple annunciator using an incandescent lamp on the sender unit and a CDS or cadmium sulfide cell as the receiver or detection unit. The *electric eye*, as it was called, now contains pulsed infrared beams that cannot be seen or easily defeated (see Fig. 5-6).

The light-beam detector can be implemented in two ways. In one method, a light-sending unit is placed in one location in line with a separate receiver or detection unit. The separate transmitter and detector are often used for long-range beam applications.

Motion sensors 139

*Fig. 5-6* *Photoelectric sensor pair.*

The second light-beam detection unit contains both transmitter and light detector in one housing, and a reflector unit bounces the light beam back to the detector cell. Wiring the self-contained light-beam detector is easier, but its range is not as great as the separate unit configuration, because the mirror reduces the usable working distance. Light-beam systems produce both normally closed and normally open outputs.

Another common alarm sensor is the *ultrasonic sensor* (see Fig. 5-7). The system contains an ultrasonic generator that excites an ultrasonic transducer. An ultrasonic signal is transmitted throughout the protected area. The ultrasonic system operates in the 19–25 kHz frequency range. A second transducer is connected to a receiver and discriminator circuit. Any differences in

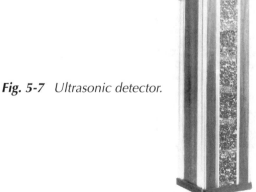

*Fig. 5-7* *Ultrasonic detector.*

the reflected signal caused by a moving object is seen as a valid trigger. The ultrasonic system is quite sensitive, but quite susceptible to false alarms, so particular attention must be paid to mounting of the ultrasonic sensor. Forced hot-air heating systems are particularly troublesome for ultrasonic systems. Moving air causes drapes or curtains to move around, thus causing false alarms. Pets also are responsible for false-alarm conditions. If you have pets in your home, avoid the ultrasonic system. Ultrasonic systems generate high-frequency audio, which bothers many people. Ultrasonic sensors are relatively low in cost and they have both normally closed and normally open contacts to provide an output. However, with the advent of passive infrared or pyroelectric sensors, ultrasonic sensors have taken a backseat.

The *pyroelectric* or *infrared body heat sensor* has become the preferred or "darling" alarm sensor, because it requires no external field excitation as used in ultrasonic systems and it is relatively free from false alarms. In the last few years, the cost of IR sensors has decreased dramatically and they have become very popular.

Pyroelectric sensors are available in many different styles and mounting configurations—from corner-mounted sensors to recessed wall-outlet type detectors (see Fig. 5-8). Passive IR sensors are also available for outdoor use as well. The outdoor units are enclosed in watertight enclosures to weather the elements. The passive IR sensor can protect over long-range patterns or in short-

***Fig. 5-8*** *Pyroelectric detector.*

range coverage. Depending upon the particular fresnel lens placed in front of the detector element, the sensor can be made to sense in different ways. In the long-range mode, the detector can protect out to 50 feet, but with a narrow beamwidth. A wide-angle lens in front of the detector allows the sensor to protect a 10×12-foot area pattern. Special lenses are available to truncate the protection beam to allow pets to walk freely in the protected area. By aiming the detector at a level mounting angle and using the special pet lens, small animals can occupy the protected space.

How the passive IR sensor is placed is important. It should never be pointed at windows that might allow sunshine to fall on the detector element. Infrared sensors should be directed at potential heat sources. Passive IR sensors are also available in portable battery-powered models. The IR sensor alarm detectors are available with normally closed and normally open outputs.

The *microwave alarm sensor* consists of an rf transmitter that operates in the 1 GHz range or above and a compatible receiver and discriminator circuit all housed in a common enclosure (see Fig. 5-9). An rf signal is sent out in the same manner as with the ultrasonic systems. The microwave sensor is generally used for indoor space protection, but it can also be used outdoors if placed in a watertight enclosure. The microwave system is sensitive to intruder detection but is susceptible to false alarms from large nearby metal objects, such as trucks or trains. When placed in metal buildings, "hot" spots can occur. The hot spots are due to bouncing reflections of rf waves inside the metal building. It is important to remember that this type of sensor generates microwaves and should not be placed in areas where people spend a considerable amount of time. Some people claim that microwave energy might be linked to cancer. If microwave sensors are used in home alarms, power should not be applied to the transmitter until the system is actually alarmed. Unfortunately, this is not usually done.

*Fig. 5-9* Microwave detector.

*Proximity sensors* have been used in the alarm industry for a long time, commonly to protect safes or file cabinets. Proximity sensors are divided into two main types—the capacitive type and the metal proximity sensor.

The *capacitive proximity detector* usually consists of a low-frequency rf oscillator in the 30 kHz range, connected to a metal pickup plate. As a person approaches the metal plate, a small amount of rf energy is bypassed to ground via the person's body, thus lowering the oscillator's frequency (see Fig. 5-10). A detector circuit senses this frequency difference and generates an alarm output. The entire proximity system is usually in a single enclosure.

***Fig. 5-10*** *Watchdog-capacitive proximity alarm.*

Another approach to the capacitive proximity detector is to use a separate rf transmitter and a separate receiver tuned to the transmitter unit's frequency. In this type of system, the intruder acts as a coupling agent between the two separate circuits. Each circuit is connected to its own metal plate. A person or large conductive object in close proximity to the transmitter plate acts as a larger radiating antenna that directs the rf energy to the receiver's antenna. Capacitive proximity detectors are generally used to protect large metal objects such as safes or file cabinets. The cabinets or safe are usually mounted on wooden blocks that lift or uncouple the object from the ground.

A variation to the capacitive proximity sensor is its simple cousin, the *capacitive touch switch*. The capacitive touch switch is usually much less sensitive than the proximity sensor. The touch switch generally uses a simple RC oscillator instead of an rf oscillator. Sometimes a CMOS chip or a 555 timer IC is used as the oscillator and detector. The touch switch acts to couple a person to the circuit via a ground path. A metal pickup plate is often used as a "spot" protection device to protect specific objects. With the touch switch, the intruder actually has to touch the object or metal plate before an output is produced. With the proximity detector, a person merely has to come near the object.

The *metal proximity detector* uses an rf oscillator to drive a tuned-capacitor/coil circuit. This type of sensor is used to detect passing vehicles, moving cars, forklifts, bicycles, etc. Metal proximity sensors are also used to sense or count metal parts on an assembly line. Proximity detectors are generally quite reliable and are available at a relatively low cost. Capacitive proximity detectors are not as commonly used as they once were; however, if you need to protect specific or sensitive materials in file cabinets during the day when most alarms are off, the capacitive proximity is a good choice. These sensors are available with both normally closed and normally open outputs.

Driveway sensors are another advance-warning or outdoor perimeter sensor intended for persons who live on large estates or have property with long driveways. There are two basic types of driveway sensors—the shallow buried air switch and the magnetic roadswitch.

The *shallow buried air switch* uses an air diaphragm switch coupled to a plastic sensing hose. The sensing hose is either liquid or air filled. As a large vehicle passes over the hose, the diaphragm switch activates and an output is coupled to the remote alarm control box. The air switch is a simple low-cost sensor. Its rather low sensitivity is actually an asset, because few false alarms are generated.

The second type of driveway switch is the *magnetic roadswitch*. This detection system uses a buried sensing coil connected to a balanced type of Maxwell bridge. When a large metal object such as a car or truck passes over the coil, the bridge circuit becomes unbalanced. A null amplifier and transistor relay driver pull in a relay that activates the alarm loop. These approximately $100 sensors are a bit more costly than the air-switch road sensor, but are quite reliable, provided the sensitivity is not adjusted for a maximum setting. The roadswitch is generally mounted near the entrance of a long or hidden driveway to give the earliest possible warning. The wires are commonly routed back to a junction box through a buried cable.

*Fence alarm sensors*, a form of vibration detector, are another type of outdoor perimeter protection. Early fence alarm sensors used a simple mechanical spring-steel lever arm with a mass attached to the free end of the lever arm. A second fixed metal contact arm is secured inside the sensor case. A set screw or sensitivity control adjusts the travel distance of the vibrating lever arm. Newer fence alarm sensors use a piezo vibration sensor. The piezo sensor uses a spring-steel arm sandwiched alongside a sheet of piezo polymer film. The sensor output is coupled to a mini IC amplifier and threshold detector circuit, all of which can be housed in a 1-×-1-×2¼-inch package. Piezo fence sensors are extremely sensitive and, therefore, use caution when adjusting sensor gain. Never set the sensitivity for maximum.

Another variation of the fence sensor uses a small coil placed underneath a small magnet attached to a vertical pendulum. When the fence moves, the pendulum moves the magnet over the coil, thus generating an output signal. The output of the pendulum detector's coil is fed to an amplifier and threshold circuit, which provides an output to drive a relay that triggers the alarm control box.

Modern vibration fence sensors are often connected to a remote discriminator or microprocessor. The discriminator acts as a time window detector that looks for a specific number of triggers in a finite time window. Microprocessors are often programmed to look for a specific "signature" signal before producing a valid output signal to the alarm control box. The major drawbacks to vibrational fence alarms are malicious children and wind.

Fenced areas can be protected in yet another way by using long-range *passive infrared telescopes*. These outdoor passive IR sensors are fitted with a special long-range lens and are mounted in outdoor enclosures. These sensors are mounted on a metal post independent of the fence itself. The sensor is then set up to look along the length of the fence. In this way, there is no fence vibration to interfere with the operation of the sensor, causing fewer false alarms. These detectors are much more expensive than the simple fence vibration sensor.

The newest long-range perimeter sensors for outdoor sensing are *laser fence alarms*. These sensors use multiple laser beams to form a criss-cross web of perimeter protection. Laser fence alarms are usually set up to look along a wall or fence. They are often used in airport security systems or to protect government installations. A laser beam is broken up into many different beams that radiate back and forth across the protected area. The outputs of sensitive pin-diode detectors are amplified and the signal is then coupled to a discriminator circuit. The multiple discriminator outputs control relay circuits to send an alarm signal back to the alarm control box.

## Smoke and fire sensors

Fire alarms are often used alongside of burglar-alarm systems. In the past, separate control boxes were often used to interface heat and smoke detectors to an alarm bell or dialer. Today, many alarm-system control boxes incorporate a second or specific channel for fire-alarm inputs.

The most commonly used low-cost sensor used in fire alarms is the *rate-of-rise* heat detector (see Fig. 5-11). This type of sensor is essentially a bimetallic-strip sensor that is sensitive to any rapid rise in ambient temperature. Rate-of-rise detectors are rated for a specific temperature range. If the room temperature rises rapidly or reaches this specific temperature, the sensor is activated. These sensors are available in both normally open or closed outputs.

*Smoke detectors* are also a very important part of an integrated fire-alarm system. Smoke detectors are divided into two

**Fig. 5-11** *Rate-of-rise heat detector.*

main classifications, the *photoelectric type sensor* and the *ionizing smoke detector*. Early fire alarms used photoelectric smoke detectors powered by 110 $V_{ac}$. A light source and a CDS photocell were combined into a collection chamber. When the light beam was disrupted by large smoke particles, the photocell's resistance changed and the detector or comparator circuit turned on the buzzer or bell.

Line-powered photoelectric smoke detectors have been eclipsed by the new ionization-type smoke detectors. Ionizing smoke detectors incorporate a minute amount of a radioactive material that ionizes the air inside of a small chamber housing the detector element. These low-cost, battery-operated sensors are very sensitive and often provide low-battery indicators and test buttons. Inexpensive ionizing smoke detectors are usually self-contained with their own buzzers, but often they do not have any provision for external alarm contacts, which are necessary to send the alarm signal back to the control box. Smoke detectors with external outputs or contacts are a little harder to find and more costly than the department-store variety (see Fig. 5-12). An avid experimenter could modify these inexpensive smoke detectors to provide a relay output. Photoelectric smoke detectors have not disappeared from the scene. Newly designed battery-powered detectors use low-current pulsed infrared LEDs and sensitive detectors to provide a new low-cost alternative to ionization detectors. Photoelectric smoke detectors are generally less sensi-

***Fig. 5-12*** *Smoke detector.*

tive to the effects of high humidity, which can be encountered when mounting smoke detectors near bathrooms.

The sensors described in this section are a representation of the most commonly used alarm sensors. Most any of the wide range of sensors shown in this book could be used to activate an alarm control box if desired. Most alarm control boxes require a normally open or normally closed contact closure to activate the control-box circuitry. The outputs from the various sensors, therefore, must be interfaced with some type of threshold detector in order to activate a relay that can trigger the alarm control box. A single comparator integrated circuit and a relay drive transistor is often all that is needed to couple the sensor to the alarm control box. Depending upon the output level of the particular sensor, an amplifier sometimes is needed between the sensor and threshold detector. Many of the circuits shown in this book already contain amplification and the needed circuitry to interface directly to an alarm control-box input.

## Cameras

Cameras are an important part of many security and surveillance systems. The two major categories include, first, nonelectronic, low-light-level-type, movie/still frame cameras, and second, electronic or television cameras. The electronic cameras are broken down into two more categories—the older vidicon cameras and the newer low-light *CCD* (charge-coupled device) cameras. CCD cameras are much more sensitive to low-light conditions. Because there is no vidicon, these cameras are much less power hungry. Only recently have CCD cameras obtained the necessary resolution to become equal to or better than vidicon cameras. CCD cameras have become very small, down to the size of a pack of cigarettes (see Fig. 5-13). They can be concealed easily behind wall-outlet-box covers.

Remote-head CCTV or closed-circuit TV cameras are now as small as 1 inch long by ½-inch diameter. These tiny remote-head cameras can be mounted just about anywhere, but they are generally more expensive than self-contained video cameras. The remote-head camera consists of the camera head unit and an electronics package coupled via an electrical cable. These tiny remote head cameras are often used with high-quality microscope imaging systems. Also available are infrared TV cameras, which are used by law enforcement agencies for nighttime surveillance. These infrared cameras are extremely sensitive but very costly.

*Fig. 5-13* Miniature low-light TV camera.

Television cameras are very useful around the home for security and surveillance applications. CCTV cameras can be used to monitor both front- and rear-door areas. When someone rings your doorbell, for example, you can immediately see what the person looks like and whether you wish the person to enter. Many security cameras are combined with an intercom, which can be used to communicate with the person at the door. A cable from the cameras are routed back to a central room in the house, such as a kitchen or den area. Usually a front- and rear-door camera are switched to a single monitor to save money. Television cameras can also be used for general monitoring around your home, whether to watch a pool area or perhaps to monitor your children in a remote playroom. A TV camera could also be combined with a VCR to monitor your children or babysitter while you are away. Wireless TV camera systems are available to send a video signal to a remote receiver monitor via an rf signal. With this arrangement, no wires are needed to link the camera to the remote monitor.

In business or industrial security systems, cameras are used to watch exits or back doors and to monitor employee pilferage in retail stores. Cameras are also used to monitor tool cribs or sensitive file areas, bank vaults, and hallways or corridors. Cameras used in these types of applications are often used with dedicated monitors and in many instances with video switchers. Video switchers concentrate a few cameras into one monitor, thus saving money. The problem with video switchers is the period of time when a particular camera is not viewed, and a well informed thief can use this to his or her advantage. Many video security systems use video switchers combined with banks of video monitors. Especially sensitive or important views are usually routed to dedicated monitors and less important scenes are routed to the switcher.

The video camera is also used with a *video motion detector*. In the early days of video monitoring, a light sensor was physically mounted or placed on the monitor screen at a specific area. To monitor a specific door, the sensor was placed over that area on the screen. As the light level changed, i.e., when a person moved through the particular scene, the light-level sensor triggered a relay to sound an alarm. In the past few years, the video motion detector has become much more sophisticated. It can monitor multiple areas at once, and they no longer have sensors mounted on the video monitor. Video motion detectors can be used to start and stop videotape recorders, which can be used later to identify the person or action of interest.

CCD cameras have become so small that it is now possible to wear a video "bug." Law-enforcement agencies can carry miniature cameras with built-in microphones to transmit both video and audio to a remote monitor and video recorder. Many "sting" operations are made possible by this high-tech aid.

Video cameras are used extensively in prisons to help corrections officers determine troubled areas or troublemakers. Small video cameras are used extensively by police. As drunk drivers are questioned, they are also directed to "walk the line," and their steps are recorded for later use. Video cameras are also used in time-motion studies to determine more efficient work procedures.

Mentioned earlier was the surveillance film camera. An 8-mm low-light film camera designed for security applications monitors employees in retail establishments. The special 8-mm film camera can be used in two ways. It can operate either as a remote-control movie camera or as a still-frame, or time-exposure, camera. The low-light film camera is less expensive than a TV-type camera used with a VCR. The film camera is used in bank ATM machines to capture still frames of each transaction.

The film camera is also used to monitor cash registers in retail stores. A switch or sensor is mounted on the cash register. Every time a cash register is opened, the camera snaps one frame per transaction. The film camera is used to monitor exit doors. After suspecting a loss or a dishonest employee, a retailer can install a magnetic reed switch to control the film camera remotely to watch rear-exit doors. Sometimes employees are found to empty the trash that also contains valuable merchandise. Rear doors are often the scene of back-door transactions. The film camera is a good psychological deterrent to discourage employees from stealing. However, the two drawbacks to film cameras include replacing batteries periodically and developing the film.

Recently, the film camera has been replaced in some instances by more expensive palm recorders. With camcorders, you no longer have to wait for film processing. The results are immediate. This approach is more costly up front, because palm recorders cost much more than film recorders, but the cost is offset by the cost of processing the film. However, film recorders are often used and the film is never processed until a theft, robbery, or ATM tampering actually occurs. Thus, money is saved if routine film is just saved or discarded without being processed.

Cameras are a valuable addition to any security or surveillance system. Costs and size of cameras continue to decline. The beauty of cameras is the fact that they can actually see, record, and identify a potential threat or an actual event.

# ❖6
# Alarm-system design philosophy

SECURITY SYSTEMS ARE AN IMPORTANT DETERRENT TO MOST would-be burglars. Persons not deterred can be scared away or apprehended. Planning and designing are the most important aspects of any security system. The first step is to assess your most vulnerable areas. Next, implement low-tech support such as lighting, locks, and window grills wherever possible. Finally, install a mix of high-tech perimeter and interior space sensors to foil any would-be thieves.

## Doors, keys, and locks

Designing a security system to protect you and your home is a multifaceted project, and it should begin with some low-tech observations. Even before you begin planning and designing your burglar-alarm system, you will need to access the weak security points around your home or office. Begin by looking at your front and back doors to ensure they are solid wooden doors. Then, look at the door between your garage and your home proper. If these doors are not solid wooden doors, they should be replaced.

Next, check your basement windows. Basement windows are extremely vulnerable, especially if surrounded by shrubs and bushes. Basement windows are easy pickings for a burglar, because they are low to the ground and often secluded. One low-cost means of protecting basement windows is to use steel bars across them. However, these bars are somewhat unsightly. Decorative steel window grills have recently become available and are available in both fixed and adjustable types. Mount them on the inside window jamb and affix them with one-way screws, if desired. Don't forget to inspect all windows and fix or replace broken window latches and/or locks. You can use some very inexpensive window locking pins to secure double-hung windows. These "pins" need only one small hole drilled into the window to secure both upper and lower window sections, and they can be fitted to allow you to keep windows open for ventilation while protecting the window against burglars.

The next step in making your home or office more secure is to check your door locks. Locks are very important for securing your property against vandals and burglars. Locks are your first line of defense against an intruder. A good lock is a good insurance policy to thwart a would-be thief. Most locks installed in new homes, apartments, and trailers are quite inadequate against an intruder. When builders construct new homes and trailers, they generally try to cut corners and, unless the prospective buyer specifies a particular lock set, the house will generally have an inexpensive lock installed. Take a look at your front and rear door locks. If you see a wedge-shaped, spring-loaded, slider mechanism with no inner shaft or slider, chances are you have an inferior lock set. Consider installing a nightlatch lock between your home and garage door, and keylock deadbolt locks above or below your existing front and rear door locks. Again, inspect your doors, and if they are not solid wood, replace them before installing new deadbolts.

The keylock deadbolt is the most secure type of lock available. It costs a little more than a regular lock, but it is money well spent. Deadbolts are positive acting locks that require a key for locking and unlocking the door. The deadbolt lock prevents the burglar from simply breaking the glass panel on the door and reaching around to unlock the door. The deadbolt lock moves a solid steel shaft in and out of the lock mechanism directly under the control of the key. There are no wedges, levers, or springs. The deadbolt lock most assuredly will slow an unsuspecting burglar and perhaps persuade him/her to move on before getting caught.

If you have recently moved into a new apartment or home, immediately consider changing the existing locks or add a deadbolt lock. It is not uncommon for past owners or renters to pick up "forgotten" items they may have left behind. Another unpleasant scenario is the unannounced return of a recently separated or divorced spouse. Sometimes renters and homeowners "forget" to surrender their old keys either by mistake or premeditated. Often, former owners kept old extra keys hidden outside and forgot them when the house was sold. It is better to be safe than sorry. Locks should not be overlooked, and you will sleep a lot easier knowing you have done everything possible to protect yourself and your family.

## Designing your alarm

When designing an alarm system, it is very important to outfox the thief. This requires extra time and considerations, but it is time well spent. When contemplating a new alarm system, take time to design your system before committing your hard-earned money. The most expensive alarm system is not worth a cent if it is improperly designed or implemented. Document your new alarm system as it is planned and executed. Documentation will greatly aid you later when servicing the system.

First, plan on using closed or "supervised" loop circuits in your new alarm. This means selecting and using normally closed switches and sensors so that any tampering or system faults will quickly be detected when the system is tested or armed. All components or sensors wired in series or "supervised" loops can quickly and reliably detect faults, which cannot be accomplished when you use open circuit-type switches and sensors wired in parallel. If you plan on using window foil and/or magnetic door and window switches, conceal the wiring and recess the door switches where possible. Wiring should not be exposed, because a wise thief can often see the wires and jumper a door switch, perhaps during the day when the alarm is inactive. Clever individuals can find an excuse to enter your home during the day, distract you, and survey your doors and wiring in an instant. The thief is more likely to be trapped once he has entered if he cannot see the sensors and the alarm wiring.

A well-placed or hidden sensor will go a long way toward trapping an unsuspecting thief. Many new sensors are wonderfully disguisable. New infrared sensors or pyroelectric sensors can mount inside a standard electrical wall-outlet box and are obscured by a cover plate.

Ultrasonic detectors can be disguised as hardcover books. New microwave sensors can be disguised as parts of furniture. Keep the burglar guessing; that's the name of the game. Don't be afraid to mix and match different types and styles of sensors. If you can afford it, use perimeter protection such as window foil and/or window pane breakage sensors and door switches along the perimeter of your home or office. Use "space protection devices such as infrared light-beam sensors, passive infrared body heat sensors, microwave sensors, or tape switch mats for your interior protection. Normally closed sensors and switches should be used, but converters are available to convert the occasional normally open-type sensor or switch to normally closed output. Splice all wiring between sensors and switches and the control box with solder. Enclose all splices in heat-shrink tubing; do not wrap with electrical tape.

Most alarm systems now use multiple loop circuits for protection. The first loop is used for perimeter protection. The second loop is used to monitor "space" or interior areas. A third loop is for advanced warning devices. Many people choose to protect their outbuildings, tool sheds, and driveways with driveway switches and/or infrared light-beam sensors. Often, another separate loop is used for certain "space" detectors within the home for nighttime protection when you are home with the alarm armed.

Most people want to use their alarm system at night when they are home, as well as for when they are away. The problem arises when people have to get up in the middle of the night to use the bathroom or to get a drink. There are several approaches for solving this dilemma. If you do not have any children, you could install a second keypad in your bedroom to disarm the system when you get up during the night. A second solution is to turn off a particular loop protecting hallway spaces around the bedroom cluster of your home. When arming your system for nighttime use, simply exclude the particular sensors or loops that protect the bedroom areas. A third approach is a delay feature to allow you extra time to turn the system off before the alarm actually sounds. You could implement this feature in conjunction with extra keypads.

Using the outdoor and indoor perimeter loops to activate a local alarm siren would scare off a possible intruder. If no further action takes place during a specified time, the system would be reset. However, if the loop was triggered again, or if an indoor "space" protection loop is triggered, the alarm could activate a telephone dialer.

## Installation and wiring tips

There are numerous tips and tricks to installing an alarm system that are generally unknown. Basement windows and air conditioners are often overlooked and, in many instances, are unprotected. Consequently, these areas become a favorite target for burglars to enter. Old basement windows generally appear to be sealed or stuck shut, but, nevertheless, are still very vulnerable to a size 10 boot. Basement windows are often located behind bushes in dark areas around your home, which is most attractive to the burglar. As mentioned earlier, these windows can be protected in a low-tech way by installing steel bars or grills across the windows' interior frames or by electronically protecting the windows via window foil and magnetic reed switches or with "pull traps."

*Pull traps* protect windows, screens, and window transoms. They consist of a metal housing that contains two ball contacts on either side of a metal separator. The separator, usually made of metal, is connected to a trip or pull wire. If the window is broken or if someone tries to open the window, the pull trap's string removes the separator and the contacts of the pull trap device are opened, thus creating an open circuit that triggers the alarm control box. The pull trap is often less costly than using window foil and magnetic reed switches. However, window foil is a good deterrent.

Air conditioners are another often-overlooked entry point. Most people feel that an air conditioner is just too heavy for someone to remove, so it is often overlooked when installing an alarm system. You can attach a magnet with double-sided foam tape or Velcro and mount the reed switch on the window jamb or wall next to the air conditioner. Use a larger magnet switch if necessary, but don't forget to secure your air conditioner.

In many instances, there is a need to protect windows left open in the summer for ventilation. This can be easily accomplished on a double-hung window. Install a second magnet about 2 to 3 inches above the first magnet on an already protected window. Simply raise the window to the second magnet opposite the reed switch. If someone moves or opens the window, the alarm will instantly trigger. The upper portion of a double-hung window can and should be protected. One method is to use another magnetic reed switch or a "pull apart" or "pull trap", because the upper window is seldom used. A "pull apart" is essentially an old-style 300-ohm TV line coupler with a molded two-pin jack and plug arrangement. The "pull apart" can be used with or without window foil. If no window foil is used, the window side of the "pull apart" is simply shorted with a small loop and attached to the window with Velcro. The other side of the "pull apart" is mounted to the window jamb and wired in series with the lower switch contacts.

When wiring windows and doors, a special door cord can be used. The door cord is a flexible cord connection used across a door or swing-out window hinge. The flexible cord with plastic molded terminal blocks at both ends allows for a two-wire circuit to be completed between the door and jamb, or between the window and the jamb. These cords are used primarily when window foil is used on a door with a window, or on swing open type windows.

When protecting oversized or loose-fitting doors, such as garage doors, a special oversized magnetic switch is used. A larger more powerful magnet and a large reed switch comprise this type of sensor. The oversized switches are generally four times as large as the common door reed switch. Window and door switches should always be connected to instantaneous loop circuits rather than delay loop circuits, which are reserved for exit/entry loops, on front and rear doors. Usually one or more special doors are selected to use with the exit/entry delay loop, to allow the homeowner up to 45 seconds to turn off his/her alarm before the alarm sounds. Many newer alarm control boxes use a digital keypad at the chosen exit/entry door to allow access. The new alarm keypads are also mounted with a prealarm buzzer to warn the alarm owner that he/she has opened the door without keying the access code. The circuit usually allows up to 45 seconds to turn off the alarm before it sounds. Any attempt to tamper with the keypad will also trigger the alarm, because most keypads also include a tamper switch. The keypads usually have two LED status indicators. The green LED indicates when the alarm loop is ready and the red LED indicates that the alarm is "armed".

## Alarm history

In the past, most people had a simple local alarm system to control a bell or siren. Early control boxes were strictly battery powered. A group of tall 1½-V alarm loop batteries were used for the single-loop protection circuit. Only as systems evolved were power supplies substituted for the loop batteries. Newer alarm systems contain a rechargeable battery pack that is kept charged by a small dc power supply, which is powered from ac line current.

Early systems were very basic. If the alarm was triggered, it sounded immediately and had to be set or turned off manually. The old alarm control boxes had no exit/entry timers or electronic siren module. This contrasts sharply with today's control boxes, which incorporate exit/entry timers, keypad control, siren modules with automatic shutoff, advanced digital controls and displays, and automatic digital telephone dialers.

Outdoor bells and sirens alone are no longer effective, especially during the daytime when most people are at work. Beat cops walking the local streets are a thing of the past. In today's sprawling suburbs, you are lucky to see a police car patrolling your neighborhood once or twice a day. As a consequence, an alarm bell or siren could ring for a long period of time before it is heard. In many communities, unattended sirens are prohibited. This brings up the topic of local alarms VS central reporting alarms.

In the past, only banks and larger businesses near police stations were connected via a direct, hardwired, leased, two-wire pair to the police alarm panel. The dedicated pair of wires reported a burglary using a supervised line that relied on polarity reversal to trigger the police alarm panel. As time passed, more and more people and smaller businesses decided that they, too, wanted to notify the police in the event of a burglary. This demand for more coverage led to the development of the automatic telephone dialer, which we will discuss later.

The fundamental question remains, "Should your alarm system make noise at all or should an alarm system be silent?" One argument states that most thefts are made by young adolescents rather than by sophisticated thieves, such as jewel thieves, and a local alarm siren would scare them away to greener pastures. The opposite argument states that if the alarm is silent, the burglar will linger and the unsuspecting thief could be caught in the act. Many experienced thieves know what they want and where to get it. So the question remains. A compromise solution, as mentioned earlier, is to use a perimeter or advanced perimeter to sound a local alarm to scare away a would-be thief or mischief maker. If no further action takes place, the alarm resets. If a space or interior sensor is activated, then the alarm would call the police immediately.

In the late 1960s and early 1970s, inexpensive telephone dialers became readily available. Everyone had the ability to report an alarm event to the police or a neighbor. The early telephone dialers were analog devices that would announce an alarm condition over a conventional dial-up phone line. These dialers are programmed with three telephone numbers. The first number is usually directed to the police, and the other numbers are for personal use. Usually, a second channel was available for fire-alarm reporting. The second channel also could dial three phone numbers. The dialer would dial each preprogrammed phone number and report the alarm condition. Once all the numbers were dialed, the dialer would reset and wait for the next alarm.

Analog telephone dialers are still currently available for less than $150 and can be purchased from a local alarm installer. They also can be found in Radio Shack stores for approximately $99. The automatic analog telephone dialer is a self contained message-reporting system that works on any private telephone line. Analog telephone dialers are also currently being used for stand-alone panic alarms and as medical and health-care reporting systems.

The control box, or alarm control panel, is the center of a burglar-alarm system. As mentioned, the control box of yesteryear was a rather simple device that consisted of a single supervised loop circuit. The control box usually had a latching relay to activate a bell or motor siren. An extra set of relay contacts sometimes activated the police alarm panel.

The next generation of alarm control boxes used transistors and SCRs for latching and sounding the alarm siren. The early transistor control boxes, unfortunately, were extremely sensitive to electrical disturbances such as radiation and electrical storms. These early transistor control boxes often called the "fox" too many times, and the police became very disturbed. Customers began to receive "bills" after a few false alarms.

False alarms should be considered a foremost matter, even today. Alarm control boxes have become much more reliable, but human error, faulty wiring, and oversensitive sensor adjustments continue to cause false alarms. Competent alarm installers will test a new alarm for a few days in a shakedown test before connecting the alarm control box to a dialer. Alarm installers have a new tool to monitor a newly installed alarm system. This tool is connected to the control box in place of the dialer. Once triggered, the system records false-alarm triggers as well as the time of day, and this system tool helps installers to locate system faults before connecting the alarm to the outside world. The police in some communities have become so upset with false alarms that customers have been charged up to $200 per false alarm.

When installing vibration sensors, pay particular attention to creating false-alarm conditions. Windows with vibration sensors move during atmospheric pressure changes and during severe electrical storms. Door slamming and children stomping can also adversely affect sensor adjustments over time. Adjusting the sensitivity of vibration sensors to maximum settings can result in many false alarms. Vibration sensors should be set for maximum and then backed down by at least 25 percent. Oversensitivity of sensor adjustments is the foremost cause of false alarms.

# High-tech control boxes

New high-tech alarm control boxes contain dedicated microprocessors. Even the simplest control boxes now contain some type of microprocessor, and lower-priced control boxes now contain many features once reserved for top-of-the-line alarm systems. Delay loops for exit/entry, 24-hour panic-switch monitoring, and siren timeout modules are now included in most lower-priced control boxes.

More expensive alarm control boxes contain and can accept keypad controls for exit/entry, and modern digital alarm control panels can bypass specific or faulty loop circuits. Each loop circuit contains an LED to help locate faults in the system. Many new alarm control boxes also provide alphanumeric displays of all operating conditions, including circuit faults and battery condition. The more expensive control boxes also provide user programming of alarm functions via an internal keypad, digital display, and internal digital telephone dialers.

An emergency-reporting panic function is incorporated in most newer control boxes. The panic-alarm function is generally active 24 hours a day, even when the main alarm is off or unarmed. In many modern alarm systems, the panic function is sometimes, but not always, used to trigger a local alarm to summon neighbors, or the panic function is sometimes used to activate a separate analog or voice dialer. Panic alarms can be wired to outdoor lamps in addition to local alarm sirens.

Modern alarm control boxes generally operate from batteries charged by a low-voltage power supply powered from the ac line. Some systems operate from a low-voltage 12-V supply. In the event of a power failure, the batteries are instantaneously switched into the system. Power failures are indicated and recorded in the alarm control box. The more advanced alarm control boxes, such as the Napco 800 and 900 series, have six to nine protective loops. New control boxes also include a flasher circuit for an outdoor strobe lamp, to get the attention of passing motorists, the police, and neighbors. Many strobe-light circuits will remain on even after the control box has been reset, and will later shut off via a separate timer. The most advanced alarm control boxes, such as the Morse MDC-16 control box, incorporates an EEPROM to allow user programming instead of the preprogrammed PROM used in most control boxes.

When installing your new alarm system, never place the control box in a conspicuous location, such as the front-door closet.

If the control box is installed in such a location, a thief could tamper or shut off the alarm system, or cut critical wires. Therefore, mount the control box in a more obscure location. When wiring the control box to sensors and switches throughout your home or office, always use stranded insulated wire instead of solid wire. Solid wire is prone to breaking after it is handled a few times. Always solder splices and connections with 60/40 nonacid solder. Cover all soldered connections with heat-shrink tubing rather than with electrical tape.

## Sirens, strobe lights, and phone dialers

Alarm control panels usually contain siren modules that automatically turn off the siren after 20 to 30 minutes. Other approaches to scaring a perpetrator include triggering a tape recorder connected to a power amplifier. The tape recorder can be programmed with an endless-loop tape containing sounds such as barking dogs or guns being fired. Given the fight-or-flight choice, most people will respond by fleeing the area immediately upon hearing these highly emotional sounds.

The last approach to burglar control consists of a solenoid-operated tear-gas canister. The perpetrator will become immobilized and sometimes is marked with a blue stain that can be used to later help identify him. A variation of this technique is to use a bright flashing strobe to blind the thief as he/she is being immobilized. Meanwhile, the telephone dialer is summoning the police. One note of caution—don't try to booby-trap your home or office, for three good reasons. First, you could hurt yourself by forgetting about the booby trap. Second, you could hurt an innocent person. Third, if you do hurt the burglar, he/she may file a lawsuit against you!

## Wireless alarm systems

Up to this point, we have only talked about hardwired sensors and control boxes. The wireless alarm system is a whole different approach to alarm installation. Wireless alarm systems have become increasingly popular in the recent past, because they are much easier to install than hardwired systems. A basic wireless alarm system consists of a wireless receiver/control box with integrated siren and several remote transmitters, some of which have built-in sensors. You can install a wireless alarm system easily in just a few hours. Because of the ease of installation, the wireless alarm system has become a preferred system. Wireless systems usually are more expensive than hardwired systems, because each protected location uses a separate alarm transmitter. The cost of transmitters, however, is offset by the intensive labor needed to install hardwired systems. If you install the alarm system yourself, the hardwired system would be cheaper.

Wireless alarm systems operate in the 300-MHz uhf radio range. The alarm transmitters all operate on the same frequency, with a range of up to 300 feet, and up to 500 feet with external antennas. The transmitter sends a 16-bit digital message, repeated up to 30 times in 2 seconds. Each transmitter's coded message is used to locate the specific sensors in more expensive wireless systems.

Most alarm transmitters have pigtail leads to connect to external sensors such as magnetic reed switches or space protection devices such as passive infrared body heat sensors. As mentioned, an inexpensive wireless system might include two transmitters, each containing a built-in reed-switch sensor to protect the door. Passive infrared sensors are now available with internal transmitters.

When installing wireless systems, you can save money by connecting a cluster of sensors to one transmitter. For example, if you wish to protect a window and an adjacent door, both door and window sensors can be connected to the same transmitter. Some installers connect additional space-protection devices to the door/window combination to monitor the interior of the home or office using a single transmitter. The alarm transmitters usually measure about 2 by 3 inches and can be mounted either with screws or by double-sided adhesive pads.

The Dicon 9000 series wireless alarm system, which retails for approximately $700, is a good representation of a modern wireless alarm system. The wireless system consists of a receiver/control box, two remote door/window sensors with transmitters, and an external siren. The system accepts various accessories such as panic transmitters, wireless smoke detectors, flood alarms, remote keypads, and, of course, more transmitters. The Dicon system offers four security zones (protection loops), and it also features a built-in telephone dialer with a voice message for security, fire, or medical emergencies. The wireless transmitters report back to the control box with specific codes that identify the transmitter location. The new wireless transmitters report back to the control box on a regular basis, thus giving "supervision" to the alarm system. When the transmitter's battery gets low (approximately one month before the end of the battery life), a trouble-light LED on the control panel lights up, instructing you to inquire via the keypad of the fault condition. The wireless transmitters use a standard 9-V transistor-radio battery. Additional transmitters are available for about $30.

The wireless receiver/control box usually operates from line voltage, but the more sophisticated systems contain a battery backup in the event of a power failure. Low-cost receiver/control boxes commonly have only a built-in siren or local alarm. The more expensive control boxes allow for external outputs, and some contain either analog or digital telephone dialers to call the police, your neighbors, or family members.

Carrier-current transmitter/receiver pairs "connect" a wireless control box and a remote siren, which can be mounted outdoors or on an attic vent. Carrier-current pairs are also used for panic buttons and to turn on remote outdoor floodlamps under the control of a wireless alarm system.

## Fire reporting

Most control boxes, whether hardwired or wireless, have some provision for fire reporting. This is an important feature that eliminates the need for a separate control box. The fire channel in an alarm control box commonly activates a second channel in a telephone dialer to call the fire department, neighbors, or family. The fire-channel input on a control box accepts smoke detectors, rate-of-rise heat sensors, and thermostats. Fire sensors or smoke detectors should also be wired in closed or supervised loops for maximum protection.

Smoke detectors are an important part of any good alarm system and they are readily available at a relatively low cost. Bargain department-store smoke detectors are usually self-contained with no external contacts to trigger an alarm control box. A skilled technician could modify these low-cost smoke detectors for use with an alarm control box. Smoke detectors with external alarm contacts are readily available, but at a higher cost than the department-store variety, from your local alarm dealer. Ionizing-type smoke detectors are, however, prone to one rather serious problem—they are very sensitive to concentrated humidity. Therefore, this type of detector should not be mounted in or near bathrooms, because shower steam will cause false alarms. Large smoke particles from greasy cooking can also often cause false alarms from sensors mounted too close to cooking surfaces.

Generally mount smoke detectors about 1 foot from the ceiling and away from any direct source of kitchen smoke. When cooking greasy foods, be sure to use a ventilation fan to reduce the likelihood of false alarms. Fortunately, many false alarms can be prevented from reaching the fire department, due to a delay feature on many new fire alarm control panels. This built-in time-delay feature allows you to abort the telephone dialer from completing the call to the fire department.

## Digital telephone dialers

Digital telephone dialers report an alarm condition to a central dispatch office instead of to the local police department. Today, many new burglar-alarm control boxes contain a built-in digital dialer instead of the once more common analog telephone dialer. Digital dialers report all alarm conditions to a central dispatch office, often hundreds of miles away. This office acts as a clearing house, or buffer, between alarm clients and local police departments. In this way, far fewer false alarms reach local police departments. Once activated, the digital dialer reports the particular alarm loop that was triggered. It indicates the location, time of day, and the names of people to contact. The dispatch office immediately ascertains the validity of the call and then calls the local police and anyone else on the contact list. Within seconds after receiving your alarm report, the local police are summoned to your home.

At the central dispatch office, operators sit in rows of monitoring alarm computers waiting for digital messages from clients around the country to log on to their system. The whole central reporting system works surprisingly well, despite the geographical distances involved. Local alarm installers usually set up the link and contact arrangements between the alarm client and the central reporting center. The alarm installer programs the digital dialer, once your alarm system has been installed. The central alarm reporting service carries a monthly charge of approximately $15 a month.

In the event of a false alarm reported to the central alarm reporting center, the alarm owner immediately enters a specific code into his alarm-panel keypad to abort the alarm message, and notifies the dispatch office of the false alarm or alarm test. If, for example, an armed intruder forced you to silence your alarm system by using the keypad, an extra number entered into your keypad tells the dispatch office that there is a possible holdup or hostage situation.

Central alarm-reporting companies have recently increased the scope of their operations. They now offer two-way video surveillance systems designed for vulnerable retail and convenience stores. The protected stores receive a silent call that automatically relays the video scenes in and around the store back to the dispatch office. These images can be and often are recorded for later use if necessary. The cameras, if used outdoors, offer an excellent deterrent against vandals. The cameras are often used with reverse and two-way audio systems to monitor the surrounding area and to announce messages to discourage loiterers from remaining around the store area. These high-tech alarm services work particularly well in high-crime areas.

Central alarm-reporting companies have become very popular with alarm installers because the installers often receive compensation for obtaining contracts between the alarm owner and the central dispatch office. The police departments like central reporting companies because they act as filters in eliminating false alarms, and they free up the police phones for more important calls. The alarm dispatch companies are a rapidly growing area of the lucrative alarm industry.

## Lighting for crime prevention

Lighting is another area of security often overlooked or taken for granted. Strictly speaking, lighting is not an alarm component, but can be a good deterrent. If you are often away from your home in the evenings, or if you frequently travel, you should consider buying an inexpensive lamp timer. The lamp timer can be connected to any table lamp in any room in your house. A dark home is inviting for a burglar; therefore, keep your home lit when you are not there. Timers are usually set to come on when it becomes dark outside and to shut off at midnight. When selecting a timer for different lamps, consider a timer with a random shutoff feature. The unpredictability of these timers is a great asset against a potential thief, because the burglar cannot determine if anyone is actually there or not.

One lamp and timer should be used in the living-room area and in at least one other area as well. Lamps with timers also keep dark areas of the home bright, such as back bedrooms or kitchen areas. Even if you set these timers to operate for only a couple of hours in the evening, they are still a valuable aid and worthwhile investment in low-cost security.

The development of passive infrared or body-heat detectors has given a great advantage to security-system designers. The IR heat sensors have become very inexpensive and are now used for lighting-control applications as well as for alarm-system sensors. Passive IR sensors are available with timers and can be integrated into standard wall-switch electrical boxes. These sensors are being used for hallway lighting. The sensors detect movement and turn on hallway lamps for an adjustable time period set by the user. These sensors assure that hall lamps will not be left on, thus wasting energy. There is also an added security feature to these lamps. If a burglar does enter your home while you are away, the security lights will follow him around the house. If you have a neighbor who is keeping tabs on your home while you are away, they could be alert for moving lights. I would love to see the look on an unsuspecting burglar's face as lights begin to follow him around the house!

Outdoor flood lighting, using passive infrared sensors, is also quite useful in deterring potential burglars. Dark areas and secluded areas around your home or property should be considered prime areas for outdoor flood lamps with IR sensors, because these areas are very attractive to potential thieves. Lighting assemblies with passive IR sensors and two flood lamps have become very inexpensive in the last few years. These lighting assemblies are often mounted in corner areas under the roof overhang, directed at backyard areas. These lighting systems have two benefits. Once an intruder enters the sensor's active zone, the lamps turn on and the potential burglar is immediately startled. Upon becoming startled, the burglar begins to wonder if there are any other surprises in store for him. Most vandals and juvenile thieves are easily deterred from this method of security. Another benefit to outdoor lighting is that neighbors would be alerted once the outdoor lamps turn on and off a few times.

Outdoor lamps benefit you when you come home to a dark area. As your car approaches and you get out of your car, the lights turn on and assist you in seeing around dark corners and help you find your keys.

Outdoor strobe lights have also become part of modern alarm systems. Bright strobe lights are usually mounted in front of your home under the overhang area. The lights are connected to the alarm control box and, once the alarm is triggered, the strobe begins to flash, thus alerting passersby, your neighbors, and the police who are looking for your home.

# Detecting bugs

Security and security systems cover a wide range of concern. Espionage and bugging are big business in the United States and internationally. In the security business, a bug is a hidden transmitter. While not directly related to burglar alarms, it is nevertheless an important security concern. Bugs are used in many different applications—from husbands watching their wives or vice versa, employers listening to employees, industrial espionage, and law enforcement. Big businesses who want the business "edge" sometimes bug their competitors who might have inside knowledge of upcoming contracts, bids, or new inventions. From a security standpoint, a bug is something which much be exterminated or removed, and there are companies solely devoted to the purpose of finding bugs, closing security leaks, and advising clients of potential problems.

The two major categories or types of bugs, included hardwired and wireless. The two distinctions of wireless bugs include battery-powered room bugs with highly sensitive microphones and telephone bugs, which are line powered by the telephone circuit. The line-powered bugs begin transmitting as soon as the telephone handset is lifted from its cradle, and cease operation when the handset is returned to the cradle. Most wireless bugs transmit in the unused portion of the FM broadcast band. The more expensive bugs operate at other frequencies in the uhf spectrum, around 300 MHz.

Radio bugs can be detected readily in two ways. One method is to use a wideband scanner with systematic scanning techniques. This technique involves listening to all frequencies in small portions of the scanner's coverage, one portion at a time, until something is heard. Another approach to detecting a radio bug is to obtain a sensitive frequency counter with a wideband preamplifier ahead of the counter, and with an antenna connected to the input of the preamplifier. In this way, a suspected area can be scanned in a short period of time. The transmitter will be located quickly and easily, and its frequency will be shown on the counter's digital display.

Hardware bugs or listening devices are very difficult and sometimes impossible to intercept, because no radiation of rf energy is present. If listening devices or hardwired bugs are suspected, look carefully for any extraneous new or sloppy wiring. Also, be aware when allowing unfamiliar or service persons into your home or office.

One often used hardwired bugging technique uses the microphone in an existing room telephone. The "hot mike" is essentially the microphone in the handset connected in such a way to send room audio back down the phone line or down an extra set or pair of wires such as the black and yellow wires. The signal is then commonly picked up in a basement or phone closet where it can be recorded on a tape recorder. You can check if your phone is bugged in this way by closing the door of the room in question, after leaving a transistor radio playing. Then identify the wires from that phone either to the basement or phone closet. Attach an audio amplifier or crystal headphone across the telephone line and across the extra black and yellow wires. If you hear the radio playing in your headphones, you can be sure that someone has bugged your phone. You can, of course, expand on this idea by checking all telephone lines. Look around for any new wires or different color wiring or any suspicious boxes or recorders. Most people who attempt to bug a room do not usually have the luxury of time to install many wires or elaborate wiring, so wireless devices often are used instead. Wireless transmitters and bugs come in all sizes and shapes, so a keen eye and ear and basic detective skills are a great asset in locating bugging devices.

# Alarm circuits and systems

THE DESIGN OF ANY COMPLEX SYSTEM USUALLY REQUIRES THAT more than one circuit be linked together to form a complete system. This chapter provides many sensing and alarm circuits, some of which can be combined or incorporated into a custom alarm control system. A few circuits are complete monitoring systems in themselves. Most alarm circuits shown can be easily combined with one another, provided the outputs of the first circuit are isolated from the second circuit, either by relays or optocouplers.

## Basic latching alarm

A basic latching alarm circuit is shown in Fig. 7-1. The circuit can be used for a wide range of simple burglar or fire alarms, as well as for portable travel alarms. The circuit incorporates both normally closed and normally open type switches. The normally closed switches could be magnetic reed switches or alarm window foil, or even relay contacts from other more complicated sensors such as light/dark detectors, pyroelectric, or solid-state vibration sensors. The normally closed switch inputs are "supervised" (alarm talk for continuous detection). Tampering or faults are quickly recognized, because all wiring is done in series. Normally, open switches or sensors are not continuous protection and are wired in parallel. A key switch could be used in parallel across one of the normally closed switches to act as an entry/exit switch to allow you to enter without triggering the alarm system.

You can mix and match a variety of sensors to form a complete fire or alarm system. The basic latching alarm can also be expanded to include a simple 555 timer chip to allow the siren or bell to remain on for a specific time period. Also, the use of a

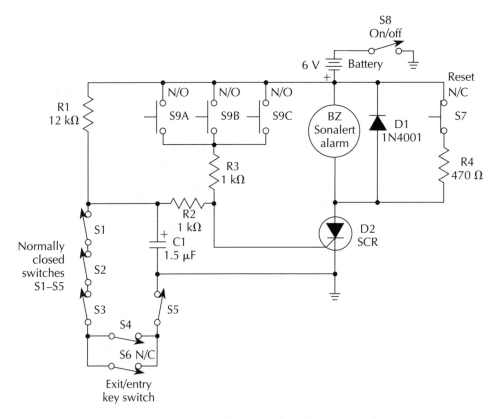

**Fig. 7-1** Basic latching alarm.

timer is recommended if you wish to control a telephone dialer. Two of the latching alarm circuits could be used to form a fire and burglar alarm system. A simple compact travel alarm could be constructed by connecting an automobile vibration sensor to the alarm circuit, and powering the alarm circuit with a 9-V battery.

## Basic latching alarm parts list

| Quantity | Part | Description |
| --- | --- | --- |
| 1 | R1 | 12-kΩ, ¼-W resistor |
| 2 | R2, R3 | 1-kΩ, ¼-W resistor |
| 1 | R4 | 470-Ω, ¼-W resistor |
| 1 | C1 | 1.5-μF, 25-V electrolytic capacitor |
| 1 | D1 | 1N4001 silicon diode |
| 1 | D2 | SCR (NTE5404) |
| 1 | BZ | Sonalert or piezo buzzer |
| 5 | S1–S5 | Normally closed sensors or switches |
| 1 | S6 | Key switch |

| 1 | S7 | Normally closed push button |
| 1 | S8 | SPST toggle switch |
| 3 | S9A, S9B, S9C | Normally open sensors or switches |
| 1 | BATT | 6-V battery |

## Remote sensing

Remote sensing or monitoring of physical phenomena such as light, temperature, and pressure can be accomplished easily by using the system shown in Fig. 7-2. The first part of the system consists of a sensor and oscillator. A sensor such as a thermistor or light-sensitive resistor is connected to a 555 timer IC configured in the stable mode. The changing sensor resistance essentially con-

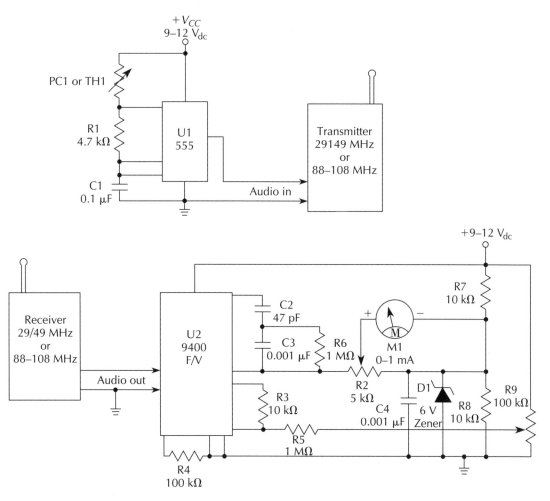

**Fig. 7-2** *Remote sensing system.*

trols the frequency of the 555 oscillator. The resultant output signal on pin 3 of the timer IC can be connected to any transmitter with an audio input. A 27/49-MHz transmitter or an FM broadcast transmitter such as a Ramsey FM-4 module could be used.

The second part of the system is the receiver and display section. Any receiver with an audio output and compatible with the transmitter frequency can be used for monitoring. An FM radio system is preferable to AM modulation, because FM is less susceptible to noise and interference. The audio output from the receiver is coupled directly to a Teledyne 9400 frequency-to-voltage converter chip. The voltage output from the 9400 converter is proportional to the frequency generated by the 555 timer in the transmitter section. The potentiometer R2 adjusts or zeros the system. The 9400 converter can be used up to 25 kHz. The output from the 9400 could be connected to an analog meter as shown, or it could be coupled to a digital panel meter or to an A/D card in a personal computer.

A variation on the remote-sensing system would be to use a tape recorder instead of a radio transmitter, as a portable data logger in the field. Couple the oscillator to the microphone or line inputs on the tape recorder. You could couple the recorder playback, when retrieved from the field, via the earphone jack to the frequency-to-voltage converter. The low-cost remote system will permit you to monitor a sensor in a remote or hard-to-reach location and to display the results in the comfort of your home or office.

## Remote-sensing system parts list

| Quantity | Part | Description |
| --- | --- | --- |
| 1 | PC1 or TH1 | 50–100-k$\Omega$ resistive photocell or thermistor |
| 1 | R1 | 4.7-k$\Omega$, ¼-W resistor |
| 1 | C1 | 0.1-$\mu$F, 25-V capacitor (disk) |
| 1 | U1 | 555 IC timer |
| 1 | TX | Miniature transmitter or walkie-talkie |
| 1 | RX | Receiver on TX frequency |
| 1 | U2 | LM9400 freq/volts converter |
| 1 | R2 | 5-k$\Omega$ potentiometer |
| 3 | R3, R7, R8 | 10-k$\Omega$, ¼-W resistor |
| 1 | R4 | 100-k$\Omega$, ¼-W resistor |
| 2 | R5, R6 | 1-M$\Omega$, ¼-W resistor |
| 1 | R9 | 100-k$\Omega$ potentiometer |

| | | |
|---|---|---|
| 1 | C2 | 47-pF capacitor (disk) |
| 2 | C3, C4 | 0.001-µF capacitor (disk) |
| 1 | D1 | 6-V zener diode |
| 1 | M | 0–1 mA meter |

## Window/door alarm

An optical window/door alarm is a compact alarm circuit that can be used in many different alarm applications, including the protection of doors, windows or pieces of equipment or antiques. When the path between the LED and phototransistor is blocked, no output is present. Once the path is opened between the LED and the detector, the output goes high and a loud piezo sounder is activated.

The optical alarm consists of two 555 timer ICS. The first timer is connected as a free-running oscillator that drives an IR LED at a frequency determined by the R1/C1 combination. A current limiter for the LED is provided by R3, as shown in Fig. 7-3, and the second 555 IC. The receive section is configured as a monostable oscillator that acts as a missing-pulse detector. The time constant of the one-shot circuit is determined by the R5/C2 combination. The monostable oscillator is looking for pulses of a specific frequency and, when missed, the alarm sounds.

An optoisolator such as a GEH1B1 can be used for short-range applications, with a small light-blocking plate between the

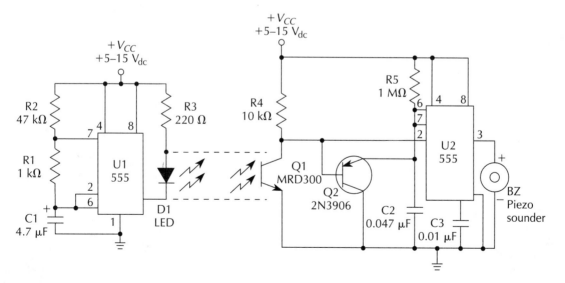

**Fig. 7-3**  *Window/door alarm.*

LED and the detector. A variation to the system is to replace the optoisolator with a separate LED and phototransistor, and place a lens in front of both the detector and LED. In this manner you can separate the LED from the detectors by tens of feet.

Another variation of the optical alarm is shown in Fig. 7-4. An LED and phototransistor are mounted near each other but angled toward each other at an angle of 30 to 45 degrees. The sensor could then detect the presence or absence of a reflective surface, i.e., white tape placed on something you wish to protect. This low-cost alarm can be adapted to many different protection or sensing applications.

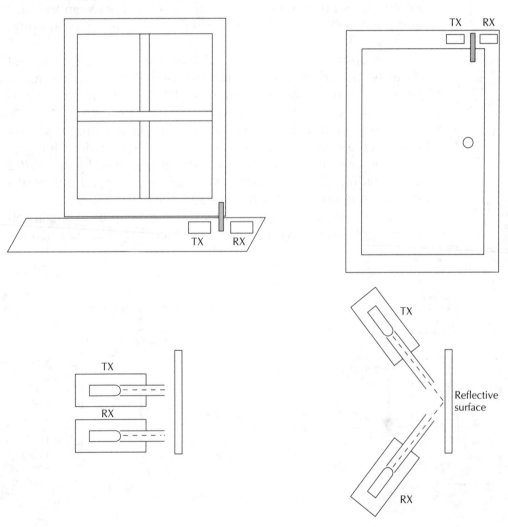

**Fig. 7-4** *Optical door/window alarm mounting.*

## Window door alarm parts list

| Quantity | Part | Description |
| --- | --- | --- |
| 1 | R1 | 1-k$\Omega$, ¼-W resistor |
| 1 | R2 | 47-k$\Omega$, ¼-W resistor |
| 1 | R3 | 220-$\Omega$, ¼-W resistor |
| 1 | R4 | 10-k$\Omega$, ¼-W resistor |
| 1 | R5 | 1-M$\Omega$, ¼-W resistor |
| 1 | C1 | 4.7-µF, 25-V electrolytic capacitor |
| 1 | C2 | 0.074-µF, 25-V capacitor (disk) |
| 1 | C3 | 0.01-µF, 25-V capacitor (disk) |
| 1 | D1 | LED |
| 1 | Q1 | Phototransistor (MRD 300) |
| 1 | Q2 | 2N3906 transistor |
| 1 | U1 | 555 IC timer |
| 1 | U2 | 555 IC timer |
| 1 | BUZ | Piezo buzzer |

# Alarm system with location display

A security alarm system generally has many sensors or switches located within a home or building complex and it is often difficult to locate a specific entry point once the alarm has been triggered. One approach to this problem is to bring all the sensor wiring back to the central alarm panel, but this is very costly both in time and materials. An alternative to this problem is to use an LM3914 dot/bar display chip as an alarm controller with *location display*. By placing different resistor values across the normally closed sensor switches, a voltage drop of a specific value would present itself when a switch contact opens in the circuit loop. The specific location would then be indicated on the LED panel, as shown in Fig. 7-5. The normally closed switches in the alarm loop could be magnetic reed switches, vibration sensors, floor mats, or a pyroelectric sensor, for example. Up to 10 switches or sensors wired in series could be used in this system, and two alarm panels could be used together to form a 20-station alarm, if desired. You could wire a key switch in parallel with one of the normally closed switches to provide an exit/entry switch so the alarm would not be triggered as you entered the protected area.

All alarm sensors are wired in series. A specific resistor value is placed across each switch (see Table 7-1), so that each switch presents a different voltage drop to the LM3914 display unit. Ten

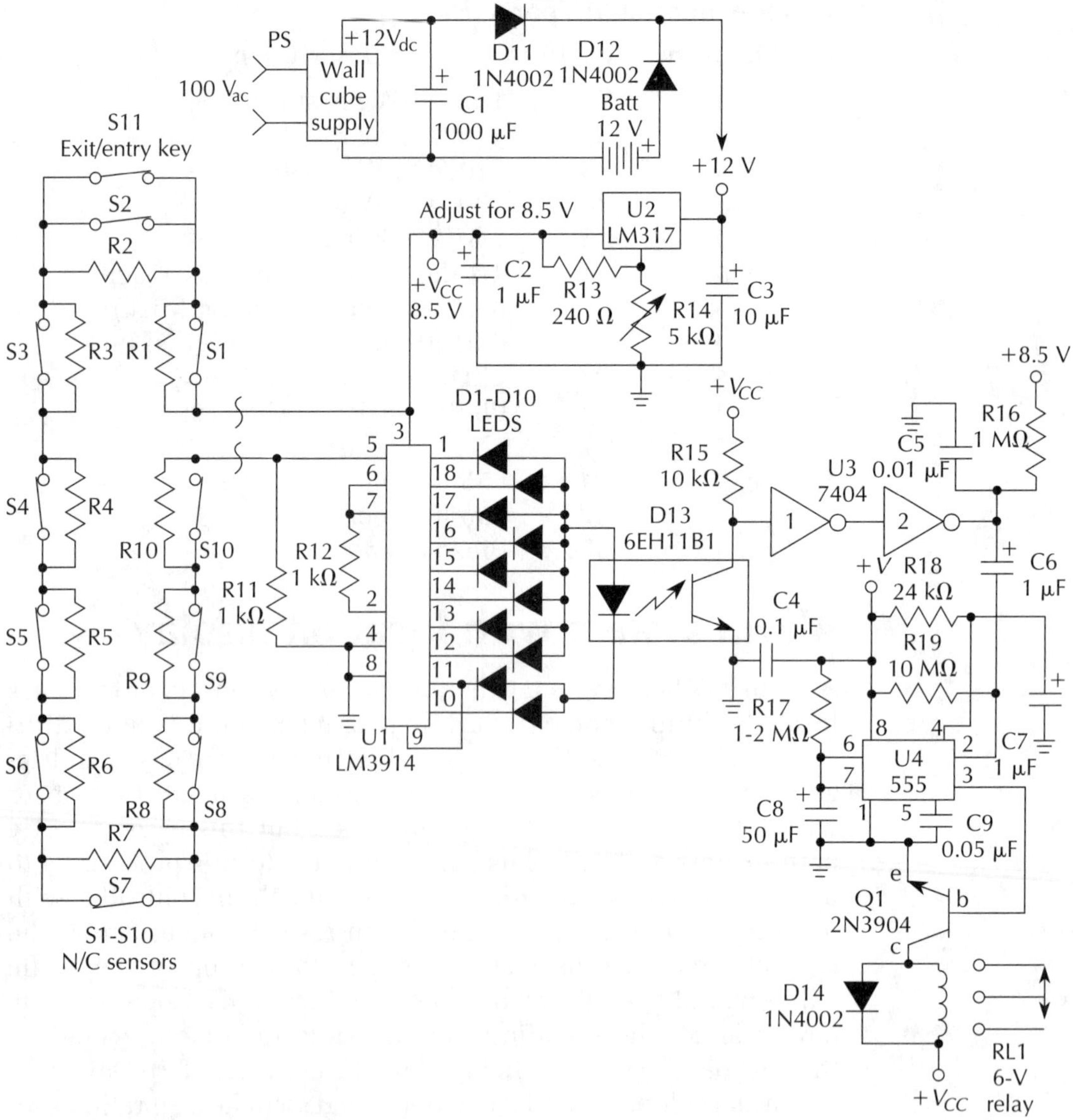

**Fig. 7-5** *Alarm system with location display.*

LEDs are wired to the LM3914 to indicate specific locations. A GEH1B1 optoisolator couples the display section to the output of the alarm system. The optocoupler drives a 555 timer IC through an inverter chip. The timer, once triggered, turns on a bell or siren for a time period determined by the R1/C1 combination. The alarm system can be powered by a 12-V gell-cell battery, which feeds an LM317 regulator set to a 8.5 $V_{dc}$. A trickle charger charges the 12-V battery.

**Table 7-1**
**Resistor selection guide**

| Resistor | Value/kΩ |
|---|---|
| R1 | 47 |
| R2 | 33 |
| R3 | 22 |
| R4 | 15 |
| R5 | 12 |
| R6 | 10 |
| R7 | 8.2 |
| R8 | 7.5 |
| R9 | 6 |
| R10 | 2–3 |

The resistors shown in Table 7-1 are typical values and should be tested at each location as the installation is completed. All switches should be recessed and the wiring should be covered so no tampering can take place. A single drawback to the location display alarm appears when two switches are simultaneously triggered. Only one lamp will light but the alarm siren will still sound. This low-cost alarm might be just what you need to protect your home or workplace.

## Alarm system with location display parts list

| Quantity | Part | Description |
|---|---|---|
| 10 | R1–R10 | See selection table (approximate values) |
| 2 | R11, R12 | 1-kΩ, ¼-W resistor |
| 1 | R13 | 240-Ω, ¼-W resistor |
| 1 | R14 | 5-kΩ potentiometer (trim) |
| 1 | R15 | 10-kΩ, ¼-W resistor |
| 1 | R16 | 1-MΩ, ¼-W resistor |
| 1 | R17 | 1–2-MΩ, ¼-W resistor |
| 1 | R18 | 24-kΩ, ¼-W resistor |
| 1 | R19 | 10-MΩ, ¼-W resistor |
| 1 | C1 | 1000-µF, 25-V electrolytic capacitor |
| 3 | C2, C6, C7 | 1-µF, 25-V electrolytic capacitor |
| 1 | C3 | 10-µF, 25-V electrolytic capacitor |
| 1 | C4 | 0.1-µF, 25-V capacitor (disk) |
| 1 | C5 | 0.01-µF, 25-V capacitor (disk) |
| 1 | C8 | 50-µF, 25-V electrolytic capacitor |

| Quantity | Part | Description |
|---|---|---|
| 1 | C9 | 0.05-µF, 25-V capacitor (disk) |
| 10 | D1–D10 | LEDs |
| 3 | D11, D12, D14 | 1N4002 silicon diode |
| 1 | D13 | GEH11B1 optocoupler |
| 1 | Q1 | 2N3904 transistor |
| 1 | U1 | LM3914 display driver (National Semiconductor) |
| 1 | U2 | LM317 regulator (National Semiconductor) |
| 1 | U3 | 7404 hex inverter |
| 1 | U4 | 555 IC timer |
| 1 | RL-1 | 6-V relay (SPST) |
| 10 | S1–S10 | Normally closed sensors or switches |
| 1 | S11 | Exit/entry key switch |

## Multipurpose alarm

An effective multipurpose alarm system is shown in Fig. 7-6. This alarm system uses both normally closed and normally open sensor inputs, which can accommodate a wide range of sensors, such as vibration sensors, infrared sensors, magnetic door switches, and window foil. The multipurpose alarm features two independent output timers. One timer controls a bell, lamp, or siren and the second timer controls an automatic dialer or a tape recorder with barking-dog sounds or gun-shot blasts on an endless-loop cassette tape, a most effective deterrent.

A 7400 NAND gate permits a supervised loop for the normally closed switches. Capacitor C1 is normally charged prior to activation of the timers. Triggering the system discharges C1 rapidly, creating a short negative pulse. Spurious spikes and noise are prevented by the integration time of R1/C1. Both timers are started simultaneously and are not retriggerable. Each timer independently controls a separate relay. The first timer is controlled by the Rt1/Ct1 combination, which activates relay RLY1. Adjust this timer for a 20-second duration to activate an automatic telephone dialer. If a tape recorder is used instead of the dialer, adjust the time period for 4 to 5 minutes. The first timer also activates a CD4013 flip-flop, which provides a system "memory" that indicates the system was triggered and can be reset by S3. The second timer is controlled by Rt2/Ct2, and Q2 drives relay RLY2 to activate a bell or siren. Adjust this second timer for a 5- to 8-minute duration. Timing for the 555 timer is

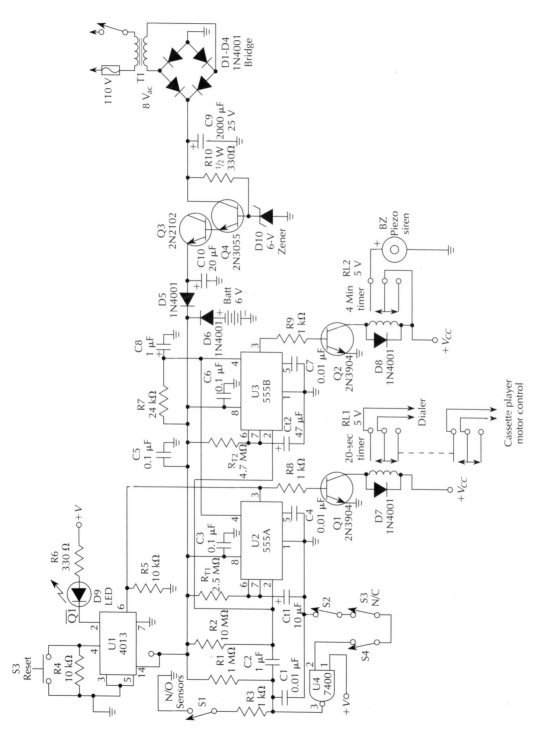

**Fig. 7-6** *Multipurpose alarm.*

determined by the formula, $T = 1.1\ R_t/C_t$, where time is in seconds, $C$ is in farads, and $R$ is in ohms.

The alarm is "fail safe" and operates even when the power fails or if someone tampers with line current. A 6-V gell-cell battery is kept charged by a charging circuit consisting of Q3 and Q4. If you plan on using a large motor siren in the system, connect an additional power relay to RLY2.

Other alarm configurations might include a dialer on the first timer and a siren on the second timer. Perhaps a solenoid could be activated by the first timer, to "pull a pin" on a tear-gas canister. The second timer could dial the police in a silent-alarm configuration. This system is quite flexible and can be configured in a number of ways, using a variety of different sensors.

## Multipurpose alarm parts list

| Quantity | Part | Description |
| --- | --- | --- |
| 1 | R1 | 1-MΩ, ¼-W resistor |
| 1 | R2 | 10-MΩ, ¼-W resistor |
| 3 | R3, R8, R9 | 1-kΩ, ¼-W resistor |
| 3 | R4, R5 | 10-kΩ, ¼-W resistor |
| 1 | R6 | 330-Ω, ¼-W resistor |
| 1 | R7 | 24-kΩ resistor |
| 1 | R10 | 330-Ω, ½-W resistor |
| 1 | Rt1 | 2.5-MΩ, ¼-W resistor or pot |
| 1 | Rt2 | 4.7-MΩ, ¼-W resistor |
| 3 | C1, C4, C7 | 0.01-μF, 25-V capacitor (disk) |
| 2 | C2, C8 | 1-μF, 25-V electrolytic capacitor |
| 3 | C3, C5, C6 | 0.1-μF, 25-V capacitor (disk) |
| 1 | C9 | 2000-μF, 25-V electrolytic capacitor |
| 1 | C10 | 20-μF, 25-V electrolytic capacitor |
| 1 | Ct1 | 10-μF, 25-V electrolytic capacitor |
| 1 | Ct2 | 47-μF, 25-V electrolytic capacitor |
| 8 | D1–D4, D5, D6, D7, D8 | 1N4001 silicon diodes |
| 1 | D9 | Red LED |
| 1 | D10 | NTE5118A 6-V zener diode |
| 2 | Q1, Q2 | 2N3904 transistor |
| 1 | Q3 | 2N2102 transistor |
| 1 | Q4 | 2N3055 transistor |
| 1 | U1 | CD4013 CMOS flip flop |
| 2 | U2, U3 | 555 timer |
| 1 | U4 | 7400 quad nand gate |

| | | |
|---|---|---|
| 1 | S1 | Normally open sensors |
| 3 | S2, S3, S4 . . . | Normally closed sensors or switches |
| 2 | RL-1, RL-2 | 6-V SPST relay |
| 1 | T1 | Transformer, 110 V |
| 1 | F1 | Lamp fuse |
| 1 | BUZ | Piezo buzzer |

## Auto burglar alarm

An auto burglar alarm is a complete, compact, inexpensive system that can be used in any vehicle, including boats and campers. The heart of the auto alarm is a 556 dual-timer IC chip. The first section, IC1A, is set up as an exit/entry delay timer configured for about 15 seconds by the C1/R3 combination, as shown in Fig. 7-7. The second 556 section, IC1B, determines how long the

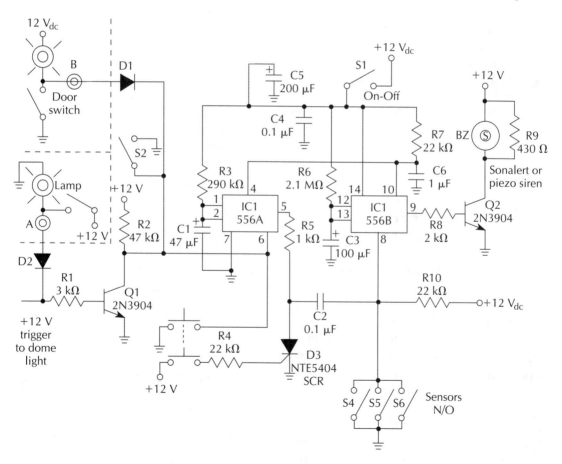

**Fig. 7-7** Auto burglar alarm.

siren or output device will sound, once triggered. This timer is controlled by the C3/R6 circuit. Pressing the DPST push-button or momentary toggle switch allows you 15 seconds to exit your vehicle. S3 triggers IC1A, and the SCR prevents IC1B from triggering during the 15-second exit period. If after 15 seconds you have not turned off the alarm via S1, the alarm will sound. Switches S4 through S6 could be replaced by vibration sensors or pyroelectric sensors and could be used inside a motor home or camper, if desired. These switches will trigger the alarm immediately, once closed.

Once triggered, the alarm remains on for a 4-minute time period, adjusted by C3/R6. The output of IC1B triggers Q2, a 2N3904 npn transistor. The transistor can drive a sonalert or piezo siren. If a larger siren is needed or if you wish to control another type of output device, the transistor could drive a larger-current relay. When driving a low-current device such as a sonalert, you might include R9 to draw some additional current. A power-on reset (POS) is formed by R7/C6 to reset pins 4 and 10 upon power-up.

The alarm can be triggered in two ways. First determine how your car doors are wired. If your door switches are wired so that 12 V is applied to your dome lamp, you can connect the door switch or dome lamp to point "A" on the alarm circuit. If your door switches activate the dome lamp by grounding the circuit via the door switch, connect to the alarm via point "B."

The circuit can be constructed on a small 2½-×-3-inch glass-epoxy circuit board placed inside a small metal box. A screw-terminal barrier strip connects the alarm circuit to the sensors, the siren, and the power source. Power for the alarm is obtained by connecting the circuit to an accessory fuse, such as the radio or stereo fuse. The siren can be mounted under the hood, facing slightly upward to prevent water from damaging the siren.

The auto burglar alarm is quite simple to install and operate. Mount S1 and S3 inside your car in a hidden place under the dash near the driver's door. Next, turn on the alarm via S1 and depress S3 when you are ready to leave your car. Anyone opening the door after the 15-second time period will activate the alarm. Activating any of the other S4 through S6 switches will also sound the alarm. The alarm circuit will also give you 15 seconds when you return to your car, to turn the alarm off via S1. The auto burglar alarm is now ready to protect your vehicle and its contents.

## Auto burglar alarm parts list

| Quantity | Part | Description |
|---|---|---|
| 1 | R1 | 3-k$\Omega$, ½-W resistor |
| 1 | R2 | 47-k$\Omega$, ¼-W resistor |
| 1 | R3 | 290-k$\Omega$, ¼-W resistor |
| 3 | R4, R7, R10 | 22-k$\Omega$, ¼-W resistor |
| 1 | R5 | 1-k$\Omega$, ¼-W resistor |
| 1 | R6 | 2-M$\Omega$, ¼-W resistor |
| 1 | R8 | 2-k$\Omega$, ¼-W resistor |
| 1 | R9 | 430-$\Omega$, ¼-W resistor |
| 1 | C1 | 4.7-$\mu$F, 25-V electrolytic capacitor |
| 2 | C2, C4 | 0.1-$\mu$F, 25-V capacitor (disk) |
| 1 | C3 | 100-$\mu$F, 25-V electrolytic capacitor |
| 1 | C5 | 200-$\mu$F, 25-V electrolytic capacitor |
| 1 | U1 | 556 dual timer |
| 2 | Q1, Q2 | 2N3904 transistor |
| 2 | D1, D2 | 1N4002 silicon diode |
| 1 | D3 | NTE5404 SCR |
| 1 | S1 | SPST on-off toggle switch |
| 1 | S2 | SPST switch |
| 1 | S3 | DPST NO push-button switch |
| 3 | S4, S5, S6 | Normally open sensors or switches |
| 1 | BUZ | Piezo buzzer |

# Bar-graph auto voltmeter

An automotive bar-graph voltmeter is an extremely useful device for any car owner. An instant glance at the bar graph displays the condition of your car battery. The LED bar-graph voltmeter also can be configured as a handy portable tester to check batteries around your home or office.

The heart of the bar-graph voltmeter is an LM3914 LED driver chip, as shown in Fig. 7-8. The minimum and maximum voltage levels can be set by the two 5-k$\Omega$ trim potentiometers.

The bar-graph meter reads from 10.5 $V_{dc}$ to 15 $V_{dc}$. The unit actually measures the 2.5 V to 3.6 V applied to the LM3914 chip. The LED driver is configured to give a dot display, in which only one of the 10 LEDs is lighted at any given time. The LEDs marked "1," "2," and "3" can be yellow instead of red, to indicate a caution condition. If the voltage of the auto battery is below 10.5 V, no LEDs will light. However, if the battery voltage is 15 V or above, the last LED, marked "10," will illuminate.

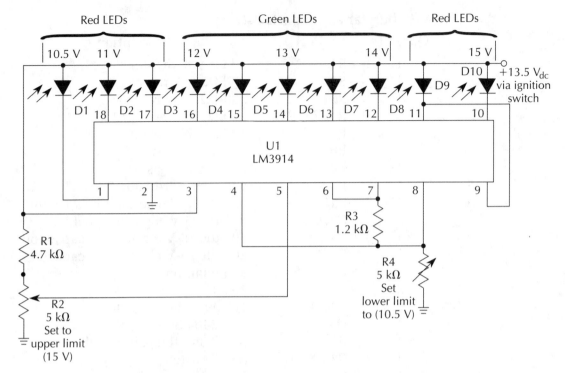

**Fig. 7-8** Bar-graph auto voltmeter.

The whole circuit can be mounted in a small plastic or aluminum chassis box. The bar-graph enclosure can be mounted under the dash as a permanent battery indicator, or it can be portable with clip leads or a cigarette-lighter plug.

### Bar-graph auto voltmeter parts list

| Quantity | Part | Description |
| --- | --- | --- |
| 1 | R1 | 4.7-k$\Omega$, ¼-W resistor |
| 1 | R2 | 5-k$\Omega$ trim pot |
| 1 | R3 | 1.2-k$\Omega$ trim pot |
| 1 | 41 | LM3914 display driver |
| 3 | D1, D2, D3 | Yellow LEDs |
| 5 | D4–D8 | Green LEDs |
| 2 | D9, D10 | Red LEDs |

## Auto immobilizer

An auto immobilizer can protect your car from unsuspecting thieves—stopping them right in their tracks. Once your car becomes immobile, the frustrated thief will most likely leave your car, rather than risk himself to an immobile, screaming car. If you own a new car, sports car, or luxury car, consider building the auto-immobilizer circuit. The auto immobilizer, shown in Fig. 7-9, was designed to protect negative-ground automobiles. Most vehicles are of this type.

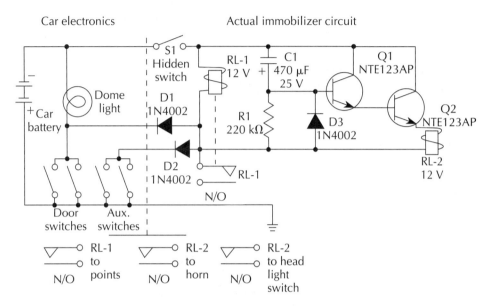

**Fig. 7-9** *Auto immobilizer circuit.*

Once the trunk or any of the doors are opened, the dome-lamp circuit applies a bias to RY-1, a DPDT relay. Relay RY-1 is immediately latched via one set of its own contacts. The second set of relay contacts shorts the car's electrical points, thus immobilizing your car. As RY-1 is being triggered, it begins a timing sequence set up by R1/C1. After this time delay, RY-2, another DPDT relay, closes. RY-2 sounds the car's horn and/or headlamps. Alternately, RY-2 could be used ahead of a flasher or interrupter module to pulse the horn and headlamp circuits, to attract attention.

Another approach to getting attention would be to connect a 12-V siren module, preferably one with a built-in timeout circuit, to one set of contacts on relay RY-2. Another approach would be to use RY-2 to short the car's electrical points. In this approach, your car would actually start but would be able to travel only a short distance before stopping.

The auto immobilizer circuit is straightforward and easy to construct. Q1 and Q2 can be surplus npn transistors to drive the

12-V relays. The heart of the auto immobilizer is the hidden switch, S1, which is used to arm and disarm the auto-immobilizer circuit, once you leave or enter your car. You could mount the hidden switch under the dash or in your glove box, or you could use an outdoor key switch instead. The immobilizer will accept other input switches or sensors. A motion sensor or sound detector could also trigger the auto immobilizer. You can mount the immobilizer circuit in a small aluminum chassis box attached to the inside firewall.

For the immobilizer to be effective, train your memory to arm and disarm your immobilizer circuits upon entering and leaving your car.

The basic auto immobilizer circuit shown in Fig. 7-9 is meant to be a starting point or inspiration for designing your custom immobilizer circuit based on your specific needs.

### Auto immobilizer circuit parts list

| Quantity | Part | Description |
|---|---|---|
| 1 | R1 | 220-k$\Omega$, ½-W resistor |
| 1 | C1 | 470-µF, 25-V electrolytic capacitor |
| 3 | D1, D2, D3 | 1N4002 silicon diodes |
| 2 | Q1, Q2 | NTE123AP transistor |
| 1 | S1 | Hidden switch or key switch |
| 2 | RL-1, RL-2 | 12-V DPDT relay |

## Digital antitheft auto lock

A digital keyless auto lock is illustrated in Fig. 7-10. You can use it to start your car or at the office for entry to sensitive areas or office locks. The heart of the keyless lock is an LS7220 four-key chip from LSI Computer Systems. This 14-pin integrated circuit is easy to implement. The circuit shown is a typical auto antitheft circuit. When the ignition switch is turned on, the sense input on pin 1 goes high and the circuit is ready to accept the unlocking sequence on pins I1, I2, I3, I4, in that order. If the input keys are pressed in sequence, I1, I2, I3, I4, but are randomly spaced and labeled, the lock control output at pin I3 will come on and the lock will energize. This state will be indicated by the off condition of the lock indicator (red LED at pin 8). An unlock condition is thus displayed. If any of the keys are depressed in any other sequence than the one described, the detector will be reset and the entire sequence must be repeated. To save the "on" condition of the lock

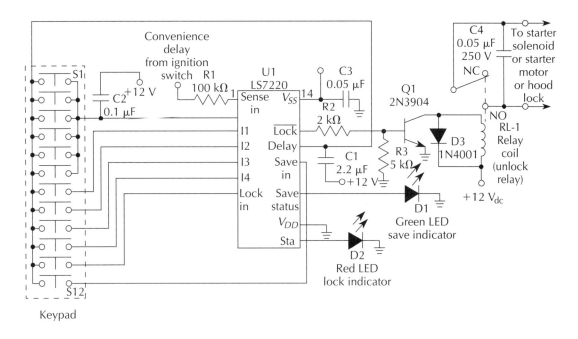

**Fig. 7-10** *Digital antitheft auto lock.*

control output when the ignition switch is turned off, i.e., when the sense becomes low, the key associated with the "save" function (pin I1) will have to be pressed. The save status is indicated by the output at pin 10 (the green LED). If the ignition switch is turned off while the green light is on, all the output status will be preserved in the chip's memory, therefore, when the ignition switch is turned on again, the input sequence is not required. This feature is ideal in parking garages or with valet parking. You can cancel "save" feature by pressing the lock key, followed by turning the ignition switch off for a time greater than the convenience delay. This will also turn off the lock-control output. The keyless lock could also be used in other applications, including safes, limited-access computer rooms, or storage areas.

### Digital antitheft auto lock parts list

| Quantity | Part | Description |
| --- | --- | --- |
| 1 | R1 | 100-kΩ, ½-W resistor |
| 1 | R2 | 2-kΩ, ¼-W resistor |
| 1 | R3 | 5-kΩ, ¼-W resistor |
| 1 | C1 | 2.2-μF, 25-V electrolytic capacitor |
| 1 | C2 | 0.1-μF, 25-V capacitor (disk) |

| Quantity | Part | Description |
|---|---|---|
| 2 | C3, C4 | 0.05-μF, 250-V capacitor |
| 1 | D1 | Green LED |
| 1 | D2 | Red LED |
| 1 | D3 | 1N4001 silicon diode |
| 1 | Q1 | 2N3904 transistor |
| 1 | U1 | LS72220 (LSI) |
| 1 | RL-1 | 12-V SPST relay |
| 12 | S1–S12 | Push buttons or keypad |

## Power-line fault detector

The novel power-line fault detector shown in Fig. 7-11 is quite useful for determining the length or duration of power spikes or

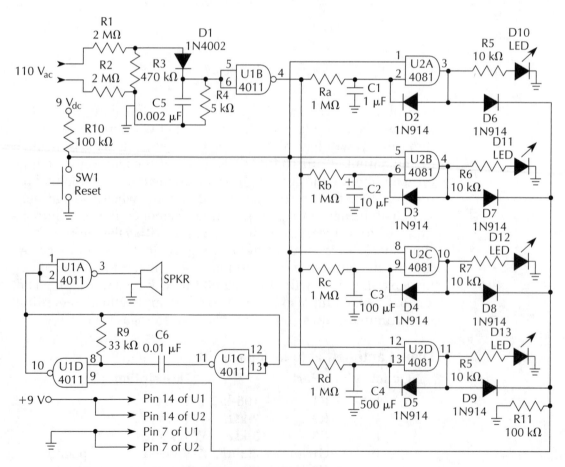

**Fig. 7-11** *Power-line fault detector.*

glitches. This greatly aids in locating the origin of the power disturbance. The ac power line is coupled to the power-line failure detector via two 2-M$\Omega$ resistors. A diode rectifies the line power and acts to provide pulse inputs to the first section of the IC. The output of the first NAND gate is forwarded to the four detector circuits, which determine the duration of the power-line disturbances. The 4081 NAND gates are set up to detect pulse intervals of 1, 10, 100, and 500 seconds. These time intervals can be altered for other pulse durations by changing resistors Ra through Rd, which are initially at 1 M$\Omega$. The RC networks are formed by resistors Ra through Rd and capacitors C1 through C4. Each detector is coupled to an LED that lights up when a pulse of a specific length is detected. All four outputs are also fed to an audio oscillator that drives an 8-$\Omega$ speaker, once an outage is detected. The speaker sounds until you push the reset button. Both integrated circuits are powered by a single 9-V battery connected to pins 7 and 14, as shown.

The power-failure detector can be mounted in a small plastic enclosure to create a portable line analyzer. This circuit could also be combined with other circuits such as a three-lamp, indicator, with neon bulbs used as a ground-wiring-fault indicator. The power-line fault detector could also activate a running-time meter or clock to help in determining exactly when the power fault occurred.

## Power-line fault detector parts list

| Quantity | Part | Description |
| --- | --- | --- |
| 2 | R1, R2 | 2-M$\Omega$, ¼-W resistor |
| 1 | R3 | 470-k$\Omega$, ¼-W resistor |
| 1 | R4 | 5-k$\Omega$, ¼-W resistor |
| 4 | R5–R8 | 10-k$\Omega$, ¼-W resistor |
| 1 | R9 | 33-k$\Omega$, ¼-W resistor |
| 1 | R10 | 100-k$\Omega$, ¼-W resistor |
| 4 | Ra, Rb, Rc, Rd | 1-M$\Omega$, ¼-W resistor |
| 1 | C1 | 1-µF, 50-V electrolytic capacitor |
| 1 | C2 | 10-µF, 50-V electrolytic capacitor |
| 1 | C3 | 100-µF, 50-V electrolytic capacitor |
| 1 | C4 | 500-µF, 50-V electrolytic capacitor |
| 1 | C5 | 0.02-µF, 100-V capacitor (disk) |
| 1 | C6 | 0.01-µF, 25-V capacitor (disk) |
| 1 | D1 | 1N4002 silicon diode |
| 8 | D2–D9 | 1N914 silicon diode |

| Quantity | Part | Description |
|---|---|---|
| 4 | D10–D13 | Red LEDs |
| 1 | U1 | CD4011 IC |
| 1 | U2 | CD4081 IC |

# Automatic emergency-lighting system

Every home or office should have some type of emergency-lighting system. The diagram in Fig. 7-12 illustrates an automatic emergency-lighting system. The circuit provides a battery-operated emergency-lighting system that is instantaneously activated if the ac line power fails. Once the line voltage is restored, the emergency light turns off and the battery is charged automatically. Emergency lighting is ideal for elevators, hallways, retail stores, or wherever the loss of light would be highly undesirable.

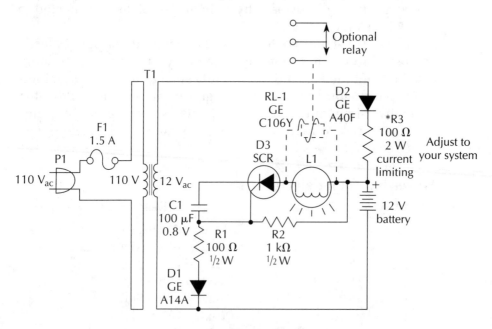

**Fig. 7-12** Automatic emergency-lighting system.

When the line voltage is normal, the capacitor charges through rectifier D1 and the 100-Ω resistor. Under these conditions, a negative voltage is provided at the gate of the SCR to ensure that the SCR is not triggered and the emergency lamp stays off. At this same time, the battery is kept fully charged by rectifier D2 through resistor R2, the current limiter. You can tailor the value of R2 to your particular battery. When the power fails, the capacitor discharges, the SCR is triggered, and the lamp turns on. The system automatically resets when the ac power is restored, because the peak ac line voltage biases the SCR and keeps it off.

You can modify the emergency-lighting system for using more than one light. You can wire a few light bulbs in parallel to serve more than one location. As you add more lamps, increase the battery size accordingly to provide more current. You could add a 6-V relay instead of a second lamp, or eliminate the lamps altogether. The relay could control larger loads or control a noisemaker to get someone's attention, or the relay could trigger an alarm panel. This simple emergency-lighting circuit can be a starting point for designing a whole emergency-lighting system throughout your home or office.

**Automatic emergency-lighting system parts list**

| Quantity | Part | Description |
| --- | --- | --- |
| 1 | R1 | 100-Ω, ½-W resistor |
| 1 | R2 | 1-kΩ, ½-W resistor |
| 1 | R3 | 100–200-Ω, 2-W current-limiting resistor |
| 1 | D1 | A14A diode (GE) |
| 1 | D2 | 40F diode (GE) |
| 1 | D3 | C106Y SCR (GE) |
| 1 | L1 | 12-V lamp |
| 1 | RL-1 | 12-V relay (optional) |
| 1 | T1 | 110-V–12-V transformer |
| 1 | BATT | 12-V gell cell |
| 1 | F1 | 1.5-A fuse |
| 1 | P1 | Power plug |

## Adjustable-rate siren

The siren is an important component in any alarm system. The adjustable rate/frequency siren shown in Fig. 7-13 is an excellent low-cost 12-V indoor siren that will frighten any intruder. The siren consists of two integrated circuits, which operate as two stable oscillators. The frequency of the LM380 oscillator is controlled by the R2/C2 combination. The LM3900 op amp acts as a stable oscillator to gate the output of the LM380 through transistor Q1. The rate control, R1 and C1, controls the rate at which the LM380 is pulsed. The output of the LM380 is coupled to a 100-μF capacitor and fed to an 8-Ω speaker via a 100-Ω level control. Pins 2, 3, and 7 of the LM380 are connected to ground.

192  Alarm Circuits and Systems

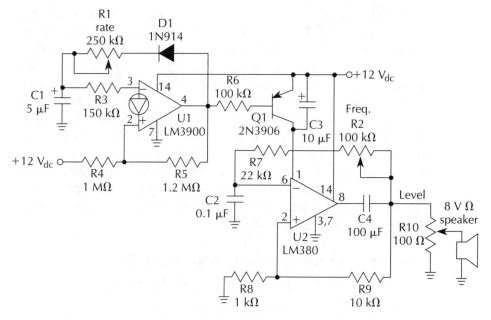

**Fig. 7-13** *Adjustable-rate siren.*

You can mount the siren in a small chassis box and power it from a 12-V battery, allowing the siren to operate in the event of a power failure or system tampering. This siren is best suited for indoor use. It most definitely will scare and/or startle the unsuspecting thief. If you need to use an outdoor siren to alert your neighbors, consider a very loud 12-V motor siren or a 20- to 30-W siren module and horn speaker. Some alarm systems use both indoor and outdoor sirens, while some people believe the best alarm is a silent alarm that notifies the police via a telephone dialer connected to the alarm control box. The alarm-philosophy section of this book (Chapter 8) discusses the pros and cons of silent alarms VS loud, noisy alarm systems.

## Adjustable-rate siren parts list

| Quantity | Part | Description |
| --- | --- | --- |
| 1 | R1 | 250-k$\Omega$ potentiometer (rate) |
| 1 | R2 | 100-k$\Omega$ potentiometer (freq) |
| 1 | R3 | 150-k$\Omega$, ¼-W resistor |
| 1 | R4 | 1-M$\Omega$, ¼-W resistor |
| 1 | R5 | 1.2-M$\Omega$, ¼-W resistor |
| 1 | R6 | 100-k$\Omega$, ¼-W resistor |
| 1 | R7 | 22-k$\Omega$, ¼-W resistor |

| | | |
|---|---|---|
| 1 | R8 | 1-kΩ, ¼-W resistor |
| 1 | R9 | 10-kΩ, ¼-W resistor |
| 1 | R10 | 100-Ω potentiometer (level) |
| 1 | C1 | 5-μF, 25-V electrolytic capacitor |
| 1 | C2 | 0.1-μF, 25-V capacitor (disk) |
| 1 | C3 | 10-μF, 25-V electrolytic capacitor |
| 1 | C4 | 100-μF, 25-V electrolytic capacitor |
| 1 | D1 | 1N914 silicon diode |
| 1 | Q1 | 2N3906 transistor |
| 1 | U1 | LM3900 op amp (National Semiconductor) |
| 1 | U2 | LM380 audio amplifier (National Semiconductor) |
| 1 | SPKR | 8-Ω speaker |

## Alarm strobe flasher

The alarm strobe flasher circuit is a useful, low-cost aid to any alarm system. The strobe-light circuit shown in Fig. 7-14 can be used in two different configurations in both indoor and outdoor applications. In an interior application, when the alarm control box is triggered, it turns on the strobe-lamp circuits along with a loud horn or siren. This combination of bright flashing lights and loud siren is sure to surprise any intruder. The strobe circuit could also be used outdoors to help the police quickly locate

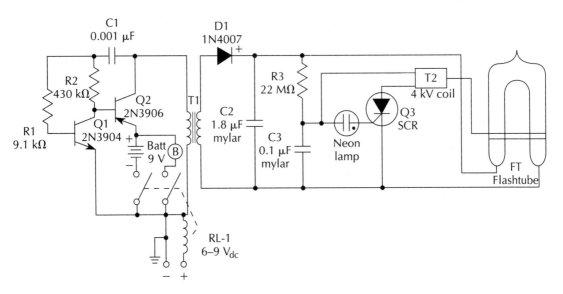

***Fig. 7-14*** *Alarm strobe flasher.*

your house and to alert your neighbors that your alarm has been activated. The strobe light makes locating or identifying your home very simple.

The alarm strobe circuit employs a direct-coupled oscillator made up of transistors Q1 and Q2 and all components to the left of transformer T1. The oscillator rapidly changes the direction of current through transformer T1. T1 is a step-up transformer consisting of a small number of primary turns and a secondary transformer of many turns. The secondary side of the transformer provides over 100 V. This high voltage is rectified and applied to capacitor C2, a 1.8-µF capacitor. A neon relaxation oscillator in the secondary circuit triggers the SCR, Q3, which, in turn, causes C3 to discharge through coil T2. As soon as the voltage applied to the neon bulb is high enough to cause the lamp to ionize, the charge of C3 is applied to T2 through the SCR. The trigger coil converts the 100 plus V from T1 to a 4,000-V pulse needed to ionize the xenon gas strobe tube. When the gas in the xenon tube is ionized, a brilliant flash is emitted. The cycle then repeats itself over and over again to keep the strobe flashing.

The strobe frequency is controlled by R3/C3. The strobe flasher can be operated from any 9- to 12-V source, including batteries. The circuit components are not particularly critical except for transformers T1 and T2, which should be readily available. You can build the circuit from scratch as a kit from the Electronic Gold Mine. The alarm strobe flasher is started by applying 6 to 9 V to the coil of a DPDT relay, as shown. You could use an SPST relay instead if you use the strobe without a noisemaker. The relay circuit isolates the flasher from the control box and makes wiring easier and straightforward.

## Alarm strobe light flasher parts list

| Quantity | Part | Description |
|---|---|---|
| 1 | R1 | 9.1-kΩ, ¼-W resistor |
| 1 | R2 | 430-kΩ, ¼-W resistor |
| 1 | R3 | 22-MΩ, ½-W resistor |
| 1 | C1 | 0.001-µF, 25-V capacitor (disk) |
| 1 | C2 | 1.8-µF, 200-V electrolytic capacitor |
| 1 | C3 | 0.1-µF, 200-V mylar capacitor |
| 1 | D1 | 1N4007 silicon diode |
| 1 | Q1 | 2N3904 npn transistor |
| 1 | Q2 | 2N3906 pnp transistor |
| 1 | Q3 | NTE5408 SCR |

| 1 | T1   | 100-V–6-V miniature transformer |
| 1 | T2   | 4-kV trigger-pulse transformer |
| 1 | NL   | Neon lamp |
| 1 | FT   | Flash tube |
| 1 | RL-1 | 6-V relay DPDT |
| 1 | B    | Piezo buzzer or siren |
| 1 | BATT | 6-V battery |

## Telephone line monitor

Most people have a number of extension phones on a single private phone line. Quite often it is difficult to tell if the telephone line is busy prior to lifting the handset, or if someone is listening while you are on the phone. The telephone line monitor in Fig. 7-15 can help you solve this problem and ensures that no one will listen to your conversations on an extension phone.

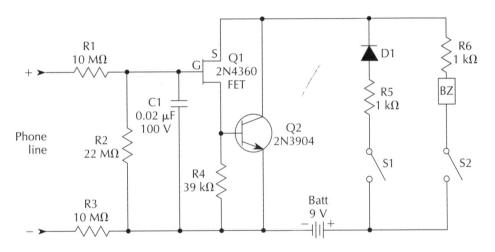

*Fig. 7-15* Phone-line monitor.

The telephone line monitor senses the voltage across the telephone line at all times. Normally, when the phone is "on hook," the voltage across the lines is about 48 $V_{ac}$. When the phone is "off hook" or picked up, the line voltage drops to about 6 $V_{dc}$. The telephone monitor is connected across the phone, i.e., in parallel via two 10-MΩ resistors. The plus (tip) lead of the phone line is connected to the FET input, as shown. The 22-MΩ resistor completes a dc current path and supplies a signal to the FET. The 0.02-μF capacitor eliminates any possible noise on the input of the FET. When all phones are "on hook," the FET's gate is biased off with a positive voltage. When the phone is lifted or "off hook," the phone-line voltage drops and the FET turns on, thus supplying current to trigger the 2N3904 npn transistor.

The telephone line monitor can drive two output devices. S1 controls the LED output and S2 controls the secondary output, which you can connect to an electronic buzzer or relay. You could use a relay to trigger another circuit or a sounder, or perhaps to start a recorder. The telephone line monitor can be powered by a standard 9-V battery, because the current consumption is very low until the output devices are activated.

To install the telephone line monitor to your phone line, first determine the polarity of the phone line. Locate the red and green phone wires. Next, use a multimeter to determine which lead is positive. Connect the positive circuit lead to this terminal. Then connect the minus lead of the telephone-line monitor circuit to the remaining phone wire. Be careful when wiring your circuit not to short the red and green phone wires, because this may cause the phone company to sense a fault condition at their central office. You are now ready to use your telephone line monitor the next time you pick up your phone.

## Telephone line monitor parts list

| Quantity | Part | Description |
|---|---|---|
| 2 | R1, R3 | 10-M$\Omega$, ½-W resistors |
| 1 | R2 | 22-M$\Omega$, ½-W resistor |
| 1 | R4 | 39-k$\Omega$, ¼-W resistor |
| 2 | R5, R6 | 1-k$\Omega$, ¼-W resistor |
| 1 | C1 | 0.02-$\mu$F, 100-V mylar capacitor |
| 1 | D1 | LED |
| 1 | Q1 | 2N4360 FET |
| 1 | Q2 | 2N3904 transistor |
| 2 | S1, S2 | SPST toggle switches |
| 1 | BUZ | Piezo buzzer |
| 1 | BATT | 9-V battery |

# Automatic telephone recorder

The automatic telephone recorder in Fig. 7-16 is a highly useful device. As a surveillance device, the telephone recorder can record or monitor a family member or caregiver or it can record an important business call. The recorder device can also be used to take notes of a detailed conversation, to be referred to later.

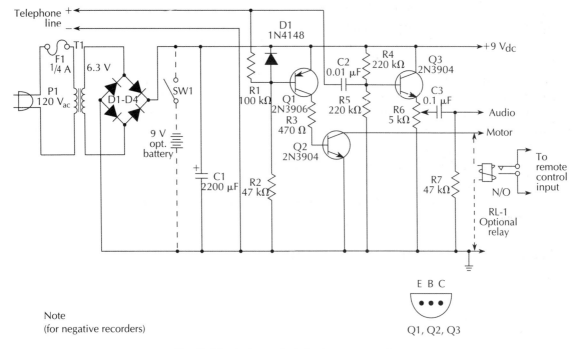

**Fig. 7-16** *Automatic telephone recorder.*

The automatic telephone recorder was initially designed to power a negative-ground tape recorder. If your portable cassette recorder is powered by a 6-V battery or it has a remote motor control jack, you could use a relay to provide switching contacts to control any type of recorder.

First, use a voltmeter to determine if your recorder has a negative ground. Connect the red lead of your meter to the plus lead of the recorder's batteries. Move the black meter lead to any point on the recorder's chassis. If your meter swings with a positive deflection, your recorder is negative ground and you can dispense with the relay.

A 6.3-V filament transformer feeds a full-wave bridge rectifier. The output of the rectifier smooths the output to a clean 9 V by capacitor C1. The 2200-μF capacitor has a large enough value to provide good filtering for most any recorder. The signal that operates Q1 and Q2 (the motor switch) is derived from the telephone line. When the telephone is "on hook," the voltage across the line is about 48 $V_{dc}$. This voltage is fed to the base of Q1 through the 47-kΩ and 100-kΩ resistors attached to Q1. This causes Q1 to be cut off, and Q2 is cut off as well. Connect the collector of Q2 to the return lead of the tape-recorder motor. The motor should be in an idle state at this point. When the phone is picked up, the line voltage drops to about 6 V. The 6-V source from the phone line forward biases through the 47-kΩ and 100-kΩ resistors to saturate Q1 into switching on. Q2 and Q3 are also biased at this time, and Q2 turns on the motor. A diode from the base to the emitter of Q1 prevents damage from the 90-V ring voltage superimposed on the telephone line. The 5-kΩ potentiometer sets the desired audio level to the input of the tape recorder.

Construct a cable from the circuit to a ⅛-inch miniature plug that is placed into the recorder's external microphone input jack. Construct another cable to power the recorder. You could use a small relay instead to activate a remote control input—even if your recorder has this type of input jack.

### Automatic telephone recorder parts list

| Quantity | Part | Description |
|---|---|---|
| 1 | R1 | 100-kΩ, ¼-W resistor |
| 1 | R2 | 47-kΩ, ¼-W resistor |
| 1 | R3 | 470-Ω, ¼-W resistor |
| 2 | R4, R5 | 220-kΩ, ¼-W resistor |
| 1 | R6 | 5-kΩ potentiometer |
| 1 | R7 | 47-kΩ, ¼-W resistor |
| 1 | C1 | 2200-µF, 25-V electrolytic capacitor |
| 1 | C2 | 0.01-µF, 25-V capacitor (disk) |
| 1 | C3 | 0.1-µF, 25-V capacitor |
| 4 | D1–D4 | 1N4002 silicon diodes |
| 1 | D5 | 1N4148 silicon diode |
| 1 | Q1 | 2N3906 pnp transistor |
| 2 | Q2, Q3 | 2N3904 npn transistor |
| 1 | R1-1 | 6–9-V SPST relay |
| 1 | T1 | 110-V–6.3-V transformer, 0.5 A |
| 1 | S1 | SPST toggle switch (if battery is used) |
| 1 | BATT | 6–9-V battery (if ac supply is not used) |
| 1 | F1 | 0.25-A fuse |
| 1 | P1 | 110-$V_{ac}$ line plug |

# rf sniffer

Security and peace of mind are very important, and locating a hidden transmitter or bug, can be very comforting. Many people are concerned about being bugged by a business partner or associates, or by a miffed spouse. The bug detector, shown in Fig.

7-17, is a very sensitive rf sniffer that can detect a 1-mV transmitter at a distance of 20 feet. This sensitive device detects the tiniest of hidden transmitters. The bug sniffer, or rf locating device, is low-cost and very effective. Frequency counters have often been used to locate hidden transmitters, but they often display random numbers caused by the frequency counter's own oscillator or other nearby transmitters. The circuit shown uses no internal oscillator; therefore, it cannot self-oscillate and generate false readings. In effect, the rf sniffer is a very sensitive rf field-strength meter using a low-cost LED bar-graph display.

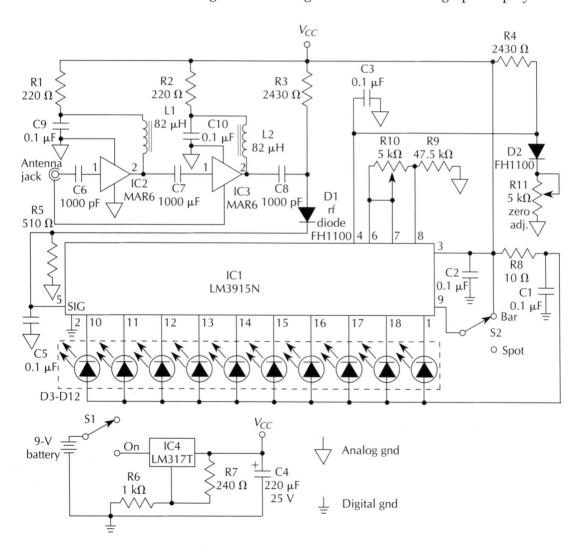

**Fig. 7-17** rf sniffer.

The rf-sniffer circuit has a two-stage wideband rf front-end amplifier and a forward-biased hot-carrier-diode detector. The detected signal is filtered and fed to IC1, an LM3915N bar-graph LED driver. The 10-segment LED display represents a 3-dB dynamic range between each successive LED. The front-end rf amplifiers are wideband devices from Mini-Circuits, P.O. Box 350166, Brooklyn, NY 11235. The monolithic microwave integrated circuit amplifiers (MMICs) 50-$\Omega$ input and output impedances from dc through 200 GHz. The gain of these amplifiers is 20 dB through 500 MHz, dropping to 11 dB at 2000 MHz. Most FM "bugs" operate in the FM broadcast band or around 300 MHz, so gain at 2000 MHz is not often a concern. The MMIC wideband amplifier modules are surface-mounted on 0.1-inch microstrip leads with surrounding surface-mounted coupling capacitors and resistors, forming a complete amplifier module. Diodes D1 and D2 are hot-carrier types, which are very sensitive to high-frequency rf.

After the detector stage, the LM3915 LED bar-graph display driver consists of a resistor divider network and a chain of op-amp comparators. The output of each comparator is open when the noninverting input is higher than the inverting input. The output goes low when the inverting input is higher. Each comparator controls a single LED on the bar graph when the comparator's output goes low. The noninverting inputs can be considered as reference inputs. Each comparator changes state as its noninverting voltage is exceeded. The LM3915 driver has an internal 1.2-V reference built in. Switch S2 programs the LM3915 for either bar or dot display. The spot or dot mode conserves power, because only one LED is on at any given time. The input to the LM3915 arrives on pin 5 as the sum of the bias voltage on the detector diode D1, plus any rectified and filtered rf from the input amplifiers IC2 and IC3. To offset the bias voltage, a low reference is generated by R4, D2, and R11. The full-scale adjustment and zero-adjust controls are both 5-k$\Omega$ potentiometers. The rf sniffer circuit can be powered by a 9-V transistor-radio battery. The battery feeds IC4, an LM317 regulator, which provides a regulated voltage for the rf sniffer circuit.

The rf sniffer should be built on a circuit board placed in a metal box to avoid stray rf pickup.

To calibrate your rf sniffer, use a low-power transmitter such as a cordless phone. Set the zero-adjust control, R10, until the leftmost LED segment is about to come on. Set the full-scale adjustment control until all segments are lit when the rf sniffer is placed next to the test transmitter. For best results, use a telescoping antenna extended to the correct length of the frequency of interest.

To sweep a room effectively for a hidden bug, first get familiar with the rf sniffer's operating characteristics in as many situations as possible. Most bugs or hidden transmitters operate in the FM broadcast band, and the more expensive FM bugs operate in the 300-MHz FM range. People pay big money to have businesses swept for hidden bugs, but you can do it yourself for less than $50 or $60. An rf sniffer kit is also available from Optoelectronics, listed in the Appendix.

**Rf sniffer parts list**

| Quantity | Part | Description |
| --- | --- | --- |
| 2 | R1, R2 | 220-Ω, ¼-W resistor |
| 2 | R3, R4 | 2.43-kΩ, ¼-W resistor |
| 1 | R5 | 510-Ω, ¼-W resistor |
| 1 | R6 | 1-kΩ, ¼-W resistor |
| 1 | R7 | 240-Ω, ¼-W resistor |
| 1 | R8 | 10-Ω, ¼-W resistor |
| 1 | R9 | 47.5-kΩ, ¼-W resistor |
| 2 | R10, R11 | 5-kΩ potentiometer |
| 6 | C1, C2, C3, C5, C9, C10 | 0.1-µF, 25-V capacitor (disk) |
| 1 | C4 | 220-µF, 25-V electrolytic capacitor |
| 2 | C6, C7, C8 | 1000-pF, 25-V capacitor (mylar) |
| 2 | D1, D2 | FH1100 diode |
| 10 | D3–D10 | LEDs |
| 2 | L1, L2 | 82-µH coil on ferrite core |
| 1 | IC1 | LM3915N display driver |
| 2 | IC2, IC3 | MAR6 op amp (MMIC Mini Circuits, Inc., P.O. Box 3501166, Brooklyn, NY 11235) |
| 1 | IC4 | LM317T regulator (optional) |
| 1 | S1 | SPST toggle switch |
| 1 | BATT | 9-V battery |

# Home guard

The home-guard alarm system shown Fig. 7-18 is a novel circuit that can protect your home or office, day or night. The home guard is a low-cost alarm dialer that can report just about any alarm condition to a friend, neighbor, or relative. You could also use the home guard to call yourself if you were vacationing or if you were just spending the day away from your relatives. One of

## 202 Alarm Circuits and Systems

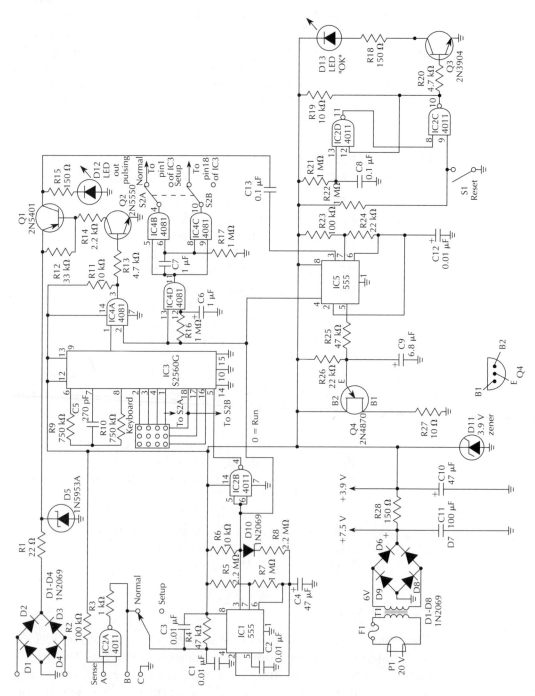

**Fig. 7-18** Home guard.

my favorite circuits, the home guard accepts most any type of alarm sensor, such as a flood sensor, freeze/thaw sensor, or people sensor. Once activated, the home guard automatically dials a preselected phone number from its memory and reports an emergency condition via a specified alarm tone.

The home guard was designed to be a simple, silent, and low-cost alarm to report specific alarm conditions such as freezing pipes or water in your basement, but it could be used for other application, as well. You can program the home guard for any telephone number and reprogram the number at any time. The home guard is transparent to your phone line and can be left connected at all times.

The heart of the home-guard circuit is IC3, an AM1 Semiconductor telephone-dialer IC. This dialer IC also incorporates a memory that holds one complete long-distance phone number. The 555 system control timer, IC1, controls the "on time"—that is, the time of dialing and waiting for the other party to answer. This timer is set for 60 seconds. The "off time" of IC1—that is, the time before retrying a busy number, is set for 30 seconds. The 555 timer is held in the standby mode from a low logic level at pin 4, during idle conditions. When the home guard is activated by a sensor signal at terminals A or B, IC1 begins its timing cycle as long as a trigger or alarm condition prevails. The output of IC1 is inverted and fed to pin 5 of IC3. This "off hook" signal tells the circuit to begin dialing its preprogrammed number. A logic-level 0 to pin 5 of IC3 causes Q1 and Q2 to be turned on, connecting LED1 to the phone line and indicating that a dial tone should be present. The signal to begin dialing, a logic-level 1, is fed to pin 1 and 18 of IC3. This signal is provided by IC4. A delay network of C9 and R22 delays the dialing by about 1 second to activate Q1. IC1 holes IC3 in the operate mode for about 1 minute. This is ample time to dial the phone number and wait for someone to answer at the opposite end of the line. When the "on time" of IC1 is completed, IC3 is returned to the "on hook" condition by the signal at pin 5. This condition disconnects the called party. Thirty seconds later, IC1 will repeat the cycle if the sensor continues to indicate a continued fault condition.

A second oscillator at IC5 generates the audio reporting tone. IC5 is only enabled during the "on time" of IC1 so that no tone is delivered to the phone line during normal use. A memory LED (LED2) indicates a memory-lost condition if power is interrupted. Loss of power is unacceptable for any alarm system. Therefore, the home-guard circuit should be powered by a ni-cad battery that is trickled-charged from the ac line. The battery would ensure that no loss of power would affect the home-guard system in the event of an actual alarm condition.

To program the number into the memory of IC3, first throw the normal/setup switch to setup. LED1 lights to indicate that the phone line is connected. Using the keypad, enter the emergency phone number of your choice. LED1 will pulse as the number is being dialed. When LED1 stops flashing, position the switch back to normal. The home guard is now ready to serve you in time of need.

Connect your sensors to the terminals marked A, B, or C. Wire a normally open sensor between terminals A and C and a normally closed sensor from B to C with a jumper between A and C.

When the home guard is initially powered up, IC2 pin 10 is held at 0 V. This is accomplished by the delay of a 0.1-µF capacitor at pin 13. Thus, Q3 goes off and the line monitor, LED2, goes dark after programming the number into IC3. Press the reset button to reactivate LED2.

The home guard is very versatile. It can be used as a burglar alarm, a freeze/thaw alarm, or a flood-alarm system. You might think of another application for the home guard, but don't forget to inform your friend or neighbor ahead of time what the tone means when they receive the call.

## Home-guard parts list

| Quantity | Part | Description |
|---|---|---|
| 1 | R1 | 22-Ω, ¼-W resistor |
| 2 | R2, R23 | 100-kΩ, ¼-W resistor |
| 1 | R3 | 1-kΩ, ¼-W resistor |
| 2 | R4, R25 | 47-kΩ, ¼-W resistor |
| 2 | R5, R8 | 2.2-MΩ, ¼-W resistor |
| 3 | R6, R11, R19 | 10-kΩ, ¼-W resistor |
| 1 | R7 | 1-MΩ, ¼-W resistor |
| 2 | R9, R10 | 750-kΩ, ¼-W resistor |
| 1 | R12 | 33-kΩ, ¼-W resistor |
| 2 | R13, R20 | 4.7-kΩ, ¼-W resistor |
| 1 | R14 | 2.2-kΩ, ¼-W resistor |
| 3 | R15, R18, R28 | 150-Ω, ¼-W resistor |
| 4 | R16, R17, R21, R22 | 1-MΩ, ¼-W resistor |
| 2 | R24, R26 | 22-kΩ, ¼-W resistor |
| 1 | R27 | 10-Ω, ¼-W resistor |
| 3 | C1, C2, C3 | 0.01-µF, 25-V capacitor (disk) |

| | | |
|---|---|---|
| 2 | C4, C10 | 47-µF, 25-V electrolytic capacitor |
| 1 | C5 | 270-pF, 25-V capacitor |
| 2 | C6, C7 | 1-µF, 25-V electrolytic capacitor |
| 3 | C8, C12, C13 | 0.1-µF, 25-V capacitor (disk) |
| 1 | C9 | 6.8-µF, 25-V electrolytic capacitor |
| 1 | C11 | 100-µF, 25-V electrolytic capacitor |
| 9 | D1–D4, D6–D9, D10 | 1N2069 silicon diodes |
| 1 | D5 | 1N5953A Zener diode |
| 1 | D11 | NTE5067A 3.9-V zener diode |
| 2 | D12, D13 | LEDs |
| 1 | Q1 | 2N5401 transistor |
| 1 | Q2 | 2N5550 transistor |
| 1 | Q3 | 2N3904 transistor |
| 1 | Q4 | 2N4870 FET |
| 2 | IC1, IC5 | 555 timer IC |
| 1 | IC2 | CD4011 CMOS IC |
| 1 | IC3 | 525606 dialer chip (AM1) |
| 1 | IC4 | CD4081 CMOS IC |
| 1 | T1 | 110-V–6.3 V transformer, 0.5 A |
| 1 | F1 | 0.5-A fuse |
| 1 | P1 | 110-V power plug |
| 1 | key | 4×3 matrix keyboard |

## Optical motion detector

The optical motion detector or balanced light detector in Fig. 7-19 is a simple, yet effective alarm sensor. If set up properly, the optical motion detector can detect humans passing tens of feet in front of its field of view. Although not as sophisticated as its cousin, the passive infrared sensor, this sensor could be used as the heart of a low-cost home alarm system. After experimenting with one of these optical motion sensors, you could devise an effective layout so that a number of these detectors could be wired together to form a complete low-cost alarm system.

The key to the optical motion detector is a surplus 6-inch-square fresnel lens placed at one end of a light-tight box. Two CDS photoresistive cells are mounted about 1 inch apart, opposite the lens, as shown. The inside of the light box is painted flat black to avoid any light or reflections of light that could interfere with the operation of the sensor. Mount the two CDS cells at the focal distance of the lens. This may prove to be the most difficult part of constructing this detector.

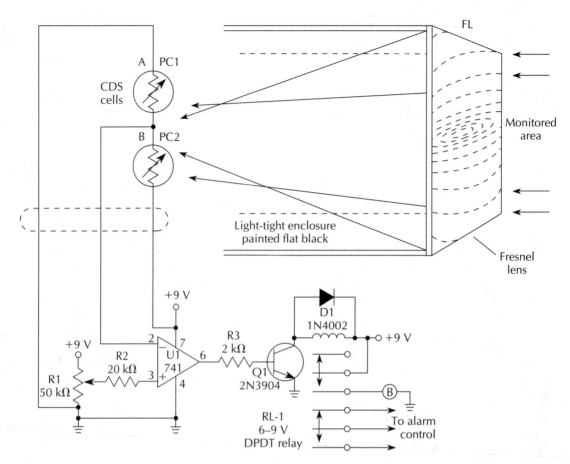

**Fig. 7-19** Optical motion detector.

The two CDS cells represent a balanced voltage to pin 2 of the 741 op amp. The 50-kΩ potentiometer is a threshold or level adjustment control. The op amp is powered by a single 9-volt battery or by the alarm control box. The 741 op amp drives Q1, a 2N3904 transistor, which powers a low-current 6–9-V relay. The DPDT relay contacts are used for two functions. One set of contacts can activate a local alarm buzzer, or it can power an LED to aid when setting up the detector. The second set of relay contacts can send an alarm signal back to a remote alarm control box.

To operate the optical motion detector, point the lens at the area to be protected and adjust the potentiometer until the relay triggers. Then back down the control until the relay drops out. Any object moving in front of the lens will imbalance the op-amp input, thus causing Q1 to activate the relay. This detector requires some background lighting in order for it to work.

## Optical motion detector parts list

| Quantity | Part | Description |
|---|---|---|
| 1 | R1 | 50-kΩ potentiometer |
| 1 | R2 | 20-kΩ, ¼-W resistor |
| 1 | R3 | 2-kΩ, ¼-W resistor |
| 2 | PC1, PC2 | 100-kΩ cadmium-sulfide photocells |
| 1 | D1 | 1N4002 silicon diode |
| 1 | Q1 | 2N3904 transistor |
| 1 | U1 | UA741 op amp |
| 1 | RL-1 | 6–9-V DPDT relay |
| 1 | BUZ | Piezo buzzer |
| 1 | FL | Fresnel lens |

## Capacitive proximity sensor

The capacitive proximity sensor shown in Fig. 7-20 is ideal for low-cost "spot" protection of specific objects around your home or office. The capacitive proximity consists of four transistors in

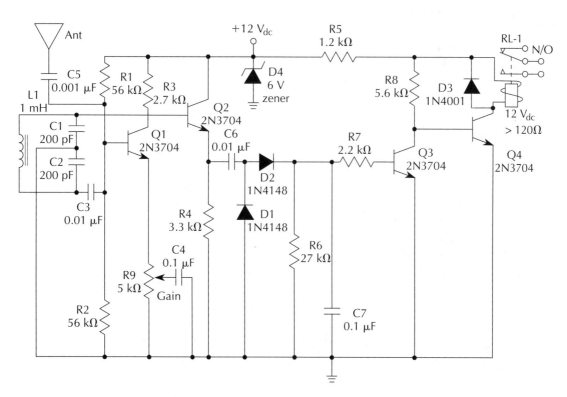

*Fig. 7-20* *Capacitive proximity detector.*

a Colpitts LC oscillator. The circuit uses capacitive loading to detect a person approaching the sensor's antenna. The gain of the rf oscillator is adjusted so that the oscillation is barely sustained. Consequently, any increase in the antenna-to-ground capacitance, such as a person approaching or touching the antenna, causes enough damping of the tank circuit to reduce the gain of the circuit below the critical value. This condition gives rise to cutting off the oscillator, thus turning on the output relay.

The proximity sensor contains a Colpitts oscillator, running at approximately 300 kHz. The antenna, which is part of the tank circuit, is coupled to the base of Q1 through capacitor C5. The gain of the circuit is adjusted by R9, a 50-k$\Omega$ potentiometer. The output of the Colpitts oscillator is buffered via Q2 and then rectified by D1 and D2 to produce a positive bias, which is fed to the base of Q3. When Q3 is operating normally (in the idle state), it is driven to saturation and Q4 is kept cut off, keeping the relay off. When the antenna is approached or touched, Q1 is cut off, which turns off Q3. At this point, Q4 is positively biased via R8 and the relay pulls in, sounding the alarm. Transistors Q1 and Q2 are fed with 6 $V_{dc}$ via diode D4, thus ensuring good oscillator stability. All transistors are common low-cost npn types. The relay should have a resistance of greater than 120 $\Omega$ to prevent circuit loading. Any suitable relay could be used. A DPDT relay would be needed if a local alarm is desired. For optimum results, the ground point of the circuit, including the emitters of Q3 and Q4, should be connected to a good ground or grounding plate. The antenna could be a copper circuit board or metal plate.

Proximity sensors of this type are often used to protect metal file cabinets. The protected cabinets are connected to the sensor's antenna plate. The cabinets must be lifted from a ground potential by putting wooden blocks underneath them. Depending upon your particular application, the circuit can be powered either by your alarm control box or from a small battery charged from a dc wall-cube power supply. The proximity detector can detect a person up to 10 inches away, depending upon the configuration uses. The capacitive proximity sensor would make an excellent addition or starting point for your custom alarm system.

## Capacitive proximity detector parts list

| Quantity | Part | Description |
| --- | --- | --- |
| 2 | R1, R2 | 56-k$\Omega$, ¼-W resistor |
| 1 | R3 | 2.7-k$\Omega$, ¼-W resistor |

| | | |
|---|---|---|
| 1 | R4 | 3.3-kΩ, ¼-W resistor |
| 1 | R5 | 1.2-kΩ, ¼-W resistor |
| 1 | R6 | 27-kΩ, ¼-W resistor |
| 1 | R7 | 2.2-kΩ, ¼-W resistor |
| 1 | R8 | 5.6-kΩ, ¼-W resistor |
| 1 | R9 | 5-kΩ potentiometer |
| 2 | C1, C2 | 200-pF, 25-V capacitor |
| 2 | C3, C6 | 0.01-µF, 25-V capacitor (disk) |
| 2 | C4, C7 | 0.1-µF, 25-V capacitor |
| 1 | C5 | 0.001-µF, 25-V capacitor (disk) |
| 2 | D1, D2 | 1N4148 silicon diode |
| 1 | D3 | 1N4001 silicon diode |
| 1 | D4 | NTE5070A 6-V zener diode |
| 1 | L1 | 1-mH coil on ferrite core |
| 4 | Q1-Q4 | 2N3704 transistors |
| 1 | RLY-1 | 12-V SPDT relay |
| 1 | ANT | Telescoping whip or wire antenna |

## Microwave motion detector

The Doppler microwave system shown in Fig. 7-21 is a self-contained, highly sensitive, motion detector. The microwave motion sensor can be used for intrusion detection or for motion-sensing applications to detect the speed of moving objects. The microwave transmitter and receiver combination forms a true Doppler detection system.

As sound, light, or radio waves reflect back from a moving object, the frequency of these waves increases as the object moves toward you, and decreases as the object moves away. You can observe this Doppler effect by listening to a train whistle as the train moves past you. The microwave sensor system incorporates both a transmitter and receiver station. The transmitter operates at about 1 GHz. It radiates rf energy in an omnidirectional pattern by means of a short wire antenna. The oscillator's frequency is determined by an etched PC stripline and the oscillator components near D1 and Q3, which include R3, C7, R7, etc. If an object moves within the radiated area, the rf waves will be reflected back to the antenna with either a higher or lower frequency. The circuit doesn't care about direction of the object, but only that it has moved. The radiated and reflected signals are mixed at diode D1. The difference between these frequencies is between 10 Hz and 40 Hz.

210  Alarm Circuits and Systems

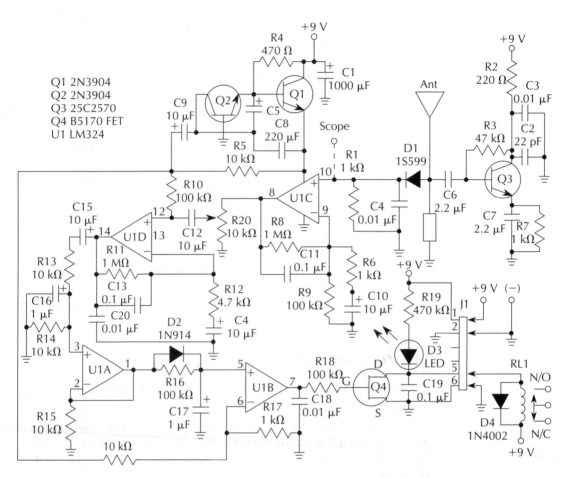

*Fig. 7-21* Microwave motion detector.

The four-section LM324 op amp consists of a receiver, amplifier, bandpass filter, and relay driver. Section U1-C of the LM324 is used as the first amplifier after detector diode D1. This amplifier feeds R8, the sensitivity potentiometer. Section U1-D functions as a bandpass filter and amplifier. The signal is then coupled to U1-A, which provides the drive signal to Q4. Transistor Q4 can drive a low-current relay. Transistors Q1 and Q2 form a precise voltage regulator to bias the LM324 amplifier.

The microwave motion sensor can be interfaced in a number of ways. The interface shown drives a relay that can activate an alarm control panel. The sensor's output could also provide a dc voltage output or it could drive an external time-delay or siren circuit. The microwave sensor can be powered by a 9-V battery or from a 12-V source from an alarm control panel.

The microwave sensor system can also detect the speed of

moving objects by connecting an oscilloscope to pin 10 of the LM324. You will then be able to see the doppler change as someone moves about in the room. For safety considerations, do not operate the microwave transmitter for extended periods of time within a 2-foot radius of a person's head.

### Microwave motion detector parts list

| Quantity | Part | Description |
| --- | --- | --- |
| 4 | R1, R6, R7, R17 | 1-k$\Omega$, ¼-W resistor |
| 1 | R2 | 220-$\Omega$, ¼-W resistor |
| 1 | R3 | 47-k$\Omega$, ¼-W resistor |
| 2 | R4, R19 | 470-k$\Omega$, ¼-W resistor |
| 4 | R5, R13, R14, R15 | 10-k$\Omega$, ¼-W resistor |
| 2 | R8, R11 | 1-M$\Omega$, ¼-W resistor |
| 4 | R9, R10, R16, R18 | 100-k$\Omega$, ¼-W resistor |
| 1 | R12 | 4.7-k$\Omega$, ¼-W resistor |
| 1 | R20 | 10-k$\Omega$ potentiometer (trim) |
| 1 | C1 | 1000-$\mu$F, 25-V electrolytic capacitor |
| 1 | C2 | 22-pF, 25-V capacitor |
| 4 | C3, C4, C18, C20 | 0.01-$\mu$F, 25-V capacitor |
| 6 | C5, C9, C10, C15, C12, C14 | 10-$\mu$F, 25-V electrolytic capacitor |
| 2 | C6, C7 | 2.2-$\mu$F, 25-V electrolytic capacitor |
| 1 | C8 | 220-$\mu$F, 25-V electrolytic capacitor |
| 3 | C11, C13, C19 | 0.1-$\mu$F, 25-V capacitor |
| 2 | C16, C17 | 1-$\mu$F, 25-V electrolytic capacitor |
| 1 | D1 | 1SS99 uhf diode |
| 1 | D2 | 1N914 silicon diode |
| 1 | D3 | LED |
| 1 | D4 | 1N4002 silicon diode |
| 2 | Q1, Q2 | 2N3904 transistor |
| 1 | Q3 | 2SC2570 high-frequency transistor |
| 1 | Q4 | BS170 FET |
| 1 | U1 | LM324 (National Semiconductor) |
| 1 | RL-1 | 9–12-V relay |
| 1 | J1 | Connector header |
| 1 | ANT | Wire or whip antenna |

## Carrier-current control systems

Alarm systems often need to control a remote siren or outdoor flood lamps. Alarm control panels, wireless receivers/control boxes, and telephone dialers are usually mounted in basements or closets near power and telephone lines. Many alarm systems incorporate a remote siren placed in attic rafters and, in many instances, outdoor flood lights are used to illuminate dark areas near the home. In remote places, it's often difficult to add additional wiring to connect external sirens or lights. One solution is the *carrier-current control system*.

The carrier control system consists of a transmitter and receiver that pass a control signal from an alarm panel to a remote siren over the existing power wires in your home or office. This FM carrier system provides excellent quality and freedom from line noise.

The transmitter section of the line-current system, in Fig. 7-22, uses a 555 tone oscillator that modulates an LM566 voltage-controlled oscillator chip. The carrier frequency of the transmitter is determined by the R4/C4 combination and is set for 200 kHz. The 555 oscillator generates a specific tone frequency of 2 kHz. The output of the 555 is coupled to the VCO via R1, a 10-k$\Omega$ trimmer. To minimize distortion, the transmitter deviation is limited to 0.15-V peak on pin 5 of the 566 chip. The output of the 566 is a frequency-modulated square wave of about 6 $V_{p-p}$ supplied through Q1. The output signal is coupled to the power line via T1, a Toko coil. Because T1 is tuned to 200 kHz, it appears as a high-impedance collector load, so Q1 does not need to have ad-

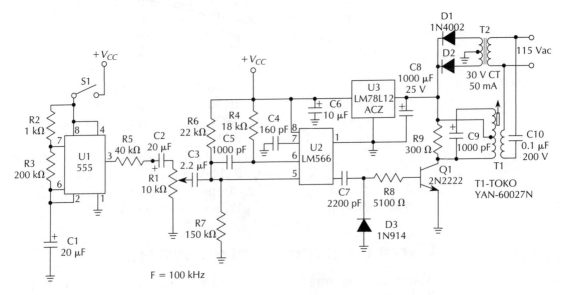

**Fig. 7-22** *Carrier-current transmitter.*

ditional current limiting. The collector signal may be as high as 40–150 $V_{p-p}$. The 0.1-μF, 200-V capacitor connected to T1 isolates the transformer from the 60-Hz line voltage.

The LM78L12 provides the necessary regulation for the VCO. Applying voltage through S1 to the 555 oscillator applies a control tone to the system carrier, which controls the receiver output relay.

The receiver section in Fig. 7-23 amplifies, limits, and demodulates the received FM signal from the transmitter section. The output of the receiver is coupled to a 2-kHz tone decoder. The tone decoder activates a control relay to apply power to a remote siren or outdoor lamps. The carrier signal is capacitively coupled from the line to tuned transformer T1, a second Toko coil. The loaded Q of the secondary of T1 and C2, a 3900-pF capacitor, is decreased by a 3.6-kΩ resistor, which limits the 10% modulated carrier and prevents ringing and spikes from entering the receiver system.

The secondary of T1 is tapped to match the base current of Q1A. The recovered carrier at the secondary of T1 might be anywhere from 0.2 to 45 $V_{p-p}$. The base of Q1A might have signal levels from 12 mV to 2.6 V. Q1A through Q1D operate as a two-stage limiter amplifier with a symmetrical square-wave output of 7 $V_{p-p}$. The output of the limiting amplifier is coupled directly to the mute peak detector. The signal is then reduced to 1 $V_{p-p}$ for driving the phase-locked loop (PLL) chip. The LM565 PLL detector operates as a narrowband tracking filter that tracks the input signal and provides a low-distortion demodulated audio output with a high signal-to-noise ratio. The oscillator within the PLL is set to free-run at a 200-kHz carrier frequency.

A mute circuit quiets the receiver in the absence of a carrier. The mute detector consists of D1, Q2, and C7. When a carrier is present, the 7-V square wave from the limiter amplifier is peak detected, and the resultant negative output is integrated by R9/C7 averaged by R10 across C7 and further integrated. The peak-detector integration and averaging prevents noise spikes from deactivating the mute in the absence of a carrier, when the limiter-amplifier output is a series of narrow 7-V spikes. The output from the LM380 audio amplifier is fed to audio interstage transformer T3, which couples the carrier-current receiver with a tone decoder. The NE567 tone detector is set for the 2-kHz frequency, as is the transmitter's oscillator. The output on pin 8 of the 567 is connected to a 6-V relay, which can control a power relay that, in turn, controls a siren or outdoor lighting circuits.

## 214 Alarm Circuits and Systems

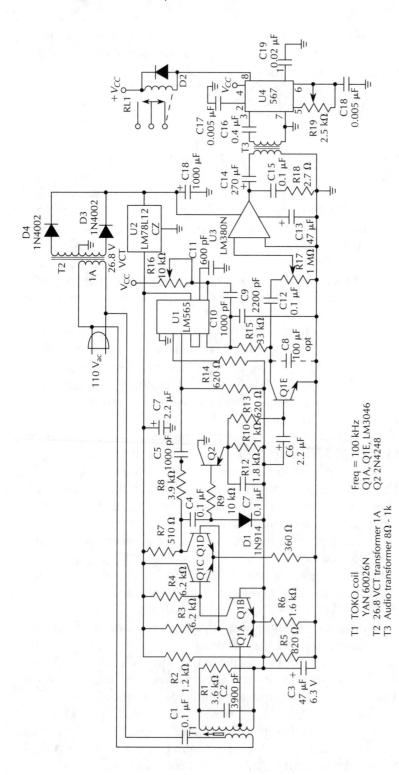

**Fig. 7-23** Carrier-current receiver.

The carrier control circuit is a complete control system that could also send a sensor signal to the input of an alarm control panel from a remote location.

## Carrier-current transmitter parts list

| Quantity | Part | Description |
|---|---|---|
| 1 | R1 | 10-k$\Omega$ potentiometer |
| 1 | R2 | 1-k$\Omega$, ¼-W resistor |
| 1 | R3 | 200-k$\Omega$, ¼-W resistor |
| 1 | R4 | 100-k$\Omega$, ¼-W resistor |
| 1 | R5 | 40-k$\Omega$, ¼-W resistor |
| 1 | R6 | 22-k$\Omega$, ¼-W resistor |
| 1 | R7 | 150-k$\Omega$, ¼-W resistor |
| 1 | R8 | 5.1-k$\Omega$, ¼-W resistor |
| 1 | R9 | 300-k$\Omega$, ¼-W resistor |
| 2 | C1, C2 | 20-$\mu$F, 25-V electrolytic capacitor |
| 1 | C3 | 2.2-$\mu$F, 25-V electrolytic capacitor |
| 1 | C4 | 160-pF, 25-V capacitor (mylar) |
| 2 | C5, C9 | 1000-pF, 25-V capacitor |
| 1 | C6 | 10-$\mu$F, 25-V electrolytic capacitor |
| 1 | C7 | 2200-pF, 25-V capacitor |
| 1 | C8 | 1000-$\mu$F, 50-V electrolytic capacitor |
| 1 | C10 | 0.1-$\mu$F, 25-V capacitor (disk) |
| 2 | D1, D2 | 1N4002 silicon diode |
| 1 | D3 | 1N914 silicon diode |
| 1 | Q1 | 2N2222 transistor |
| 1 | U1 | 555 IC timer |
| 1 | U2 | 566 VCO (National Semiconductor) |
| 1 | T1 | Yan 60027N coil (Toko) |
| 1 | T2 | 110-V–30 V CT transformer (Soma) |
| 1 | S1 | SPST toggle switch |

## Carrier-current receiver parts list

| Quantity | Part | Description |
|---|---|---|
| 1 | R1 | 3.6-k$\Omega$, ¼-W resistor |
| 1 | R2 | 1.2-k$\Omega$, ¼-W resistor |
| 2 | R3, R4 | 6.2-k$\Omega$, ¼-W resistor |
| 1 | R5 | 820-$\Omega$, ¼-W resistor |
| 1 | R6 | 1.6-k$\Omega$, ¼-W resistor |
| 1 | R7 | 510-$\Omega$, ¼-W resistor |
| 1 | R8 | 3.9-k$\Omega$, ¼-W resistor |

| Quantity | Part | Description |
|---|---|---|
| 1 | R9 | 10-kΩ, ¼-W resistor |
| 1 | R10 | 1-kΩ, ¼-W resistor |
| 1 | R11 | 360-Ω, ¼-W resistor |
| 1 | R12 | 1.8-kΩ, ¼-W resistor |
| 2 | R13, R14 | 620-Ω, ¼-W resistor |
| 1 | R15 | 33-kΩ, ¼-W resistor |
| 1 | R16 | 10-kΩ trim potentiometer |
| 1 | R17 | 1-MΩ trim potentiometer |
| 1 | R18 | 2.7-Ω, ¼-W resistor |
| 1 | R19 | 2.5-kΩ trim potentiometer |
| 1 | C1 | 0.1-µF, 200-V capacitor (disk) |
| 1 | C2 | 3900-pF, 100-V capacitor (internal to coil) |
| 1 | C3 | 4.7-µF, 25-V electrolytic capacitor |
| 4 | C4, C13, C12, C15 | 0.1-µF, 25-V capacitor (disk) |
| 2 | C5, C10 | 1000-pF, 25-V capacitor (mylar) |
| 2 | C6, C7 | 2.2-µF, 25-V electrolytic capacitor |
| 1 | C8 | 100-µF, 25-V electrolytic (optional) |
| 1 | C9 | 2200-pF, 25-V capacitor (mylar) |
| 1 | C11 | 600-pF, 25-V capacitor (mylar) |
| 1 | C13 | 47-µF, 25-V electrolytic capacitor |
| 1 | C14 | 270-µF, 25-V electrolytic capacitor |
| 1 | C16 | 0.4-µF, 25-V capacitor |
| 2 | C17, C18 | 0.005-µF, 25-V capacitor (disk) |
| 1 | C19 | 0.02-µF, 25-V capacitor (disk) |
| 1 | D1 | 1N914 silicon diode |
| 2 | D2, D3, D4 | 1N4002 silicon diode |
| 5 | Q1A–Q1E | LM3046 transistors |
| 1 | Q2 | 2N4248 transistor |
| 1 | U1 | LM565 IC (National Semiconductor) |
| 1 | U2 | LM78L12CZ regulator (National Semiconductor) |
| 1 | U3 | LM380N (National Semiconductor) |
| 1 | U4 | LM567 (National Semiconductor) |
| 1 | T1 | Yan 60026N coil (Toko) |
| 1 | T2 | 100-V–26.8-V transformer, 1 A |
| 1 | T3 | Miniature 8-Ω–1-kΩ transformer |
| 1 | RL-1 | 6-V miniature SPST relay |
| 1 | P1 | Power plug |

# Ultrasonic-sensor system

Ultrasonic sensors have been used to active alarm systems for many years. For a long time, they were the most widely used people or "space" protection sensor. They are still used today, but they have taken a backseat to the new passive infrared body-heat sensors. The ultrasonic system shown in Figs. 7-24 and 7-25 includes a separate ultrasonic transmitter and receiver operating at 23 kHz, just above the hearing range of most people.

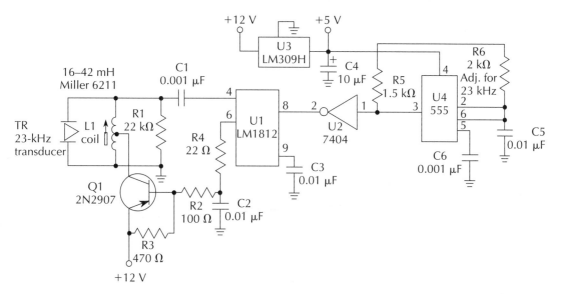

**Fig. 7-24** *Ultrasonic transmitter.*

The ultrasonic system components shown can be used in a variety of ways. You could use the transmitter separately from the receiver by mounting the transmitter on one wall and mounting the receiver on an opposite facing wall. In this way, you could use the system for long-range applications. You could house the transmitter and receiver in the same box. In this configuration, the transmitter bounces its signal off a nearby wall. This setup is best used for smaller room protection of up to 10 feet.

The ultrasonic receiver "listens" for any differences in the reflected signal caused by a person moving in the protected space. You could also use the ultrasonic transmitter and receiver as a means of remote control. By applying power to the transmitter sec-

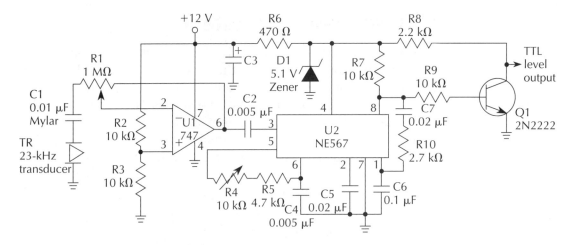

**Fig. 7-25** *Ultrasonic receiver.*

tion via S1, turning it off and on causes the receiver to detect the transmitter's audio tone. You can use the system as a panic-button transmitter or as a means of remotely controlling lights or a siren.

The transmitter section begins with an LM309 regulator, which takes the 12-V power source and regulates it down to 5 $V_{dc}$. An NE555 multipurpose timer IC is adjusted to modulate at 23 kHz. A frequency counter or oscilloscope must be used to arrive at the desired operating frequency. The output of the 555 is inverted via a 7404 and then coupled to an LM1812 ultrasonic-generator chip. The output of the inverter is directed to the modulation input of the LM1812 chip on pin 8. The output of the LM1812 drives the base of a 2N2907 transistor. The transistor drives the ultrasonic transducer.

The receive section of the system is quite straightforward. A 23-kHz ultrasonic transducer picks up the transmitter's signal. The transducer is coupled to 1-M$\Omega$ resistor to control the gain of an LM741 op amp. The output of the LM741 on pin 6 is coupled to a 0.005-$\mu$F capacitor, which feeds an NE567 tone-decoder chip. The 10-k$\Omega$ trimmer on pins 5 and 6 and the capacitor on pins 2 and 6 control the tone detector's frequency. The output of the tone detector drives a pnp transistor that provides a 5-V TTL output, which can activate a relay or alarm control box. Mount the transmitter and receiver on circuit boards placed inside a metal box, to reduce the possibility of any stray pickup and oscillations. The ultrasonic system can be powered by any 12-$V_{dc}$ source, either from the alarm control box or from a separate power supply.

When using ultrasonic systems for alarm activation, pay particular attention to placement of the sensor in relation to moving air currents. Ultrasonics have lost favor in the alarm industry mainly because they were prone to false alarms from air currents that, in turn, moved drapes or curtains, causing false alarms. Small pets or birds were often the cause of false alarms. Therefore, never place ultrasonic systems to "look" at heating ducts or drafty windows or doors. As mentioned earlier, ultrasonics can also be used as remote-control devices to activate a remote siren or as a panic-button transmitter. Ultrasonics, however, will not penetrate walls. Therefore, the panic button could only be used in a large, open room area. This is the primary reason why panic buttons are usually rf devices, because radio energy will penetrate walls and only one receiver would be necessary for an rf panic system. The ultrasonic transmitter and receiver pair can be utilized in a number of ways and the low cost of construction make it attractive for burglar-alarm use.

## Ultrasonic transmitter parts list

| Quantity | Part | Description |
| --- | --- | --- |
| 1 | R1 | 22-k$\Omega$, ¼-W resistor |
| 1 | R2 | 100-$\Omega$, ¼-W resistor |
| 1 | R3 | 470-$\Omega$, ¼-W resistor |
| 1 | R4 | 22-$\Omega$, ¼-W resistor |
| 1 | R5 | 1.5-k$\Omega$, ¼-W resistor |
| 1 | R6 | 2-k$\Omega$, ¼-W resistor (adjust value for 23kHz |
| 2 | C1, C6 | 0.001-µF, 25-V capacitor (mylar) |
| 3 | C2, C3, C5 | 0.01-µF, 25-V capacitor (disk) |
| 1 | C4 | 10-µF, 25-V electrolytic capacitor |
| 1 | Q1 | 2N2907 transistor |
| 1 | U1 | LM1812 (National Semiconductor) |
| 1 | U2 | 7404 hex inverter |
| 1 | U3 | LM309H regulator (National Semiconductor) |
| 1 | U4 | 555 IC timer |
| 1 | L1 | 16–42-mH coil (Miller 6211) |
| 1 | TR | 23-kHz ultrasonic transducer |

## Ultrasonic receiver parts list

| Quantity | Part | Description |
| --- | --- | --- |
| 1 | R1 | 1-M$\Omega$ potentiometer |
| 2 | R2, R3 | 10-k$\Omega$, ¼-W resistor |

| Quantity | Part | Description |
|---|---|---|
| 3 | R4, R9, R7 | 10-kΩ trim pot |
| 1 | R5 | 4.7-kΩ, ¼-W resistor |
| 1 | R6 | 470-Ω, ¼-W resistor |
| 1 | R8 | 2.2-kΩ, ¼-W resistor |
| 1 | R10 | 2.7-kΩ, ¼-W resistor |
| 1 | C1 | 0.01-μF, 25-V capacitor (mylar) |
| 2 | C2, C4 | 0.005-μF, 25-V capacitor (disk) |
| 1 | C3 | 100-μF, 25-V electrolytic capacitor |
| 2 | C5, C7 | 0.02-μF, 25-V capacitor (disk) |
| 1 | C6 | 0.1-μF, 25-V capacitor |
| 1 | Q1 | 2N2222 transistor |
| 1 | U1 | LM747 op amp (National Semiconductor) |
| 1 | U2 | 567 tone decoder |
| 1 | D1 | NTE135A 5.1-V zener diode |
| 1 | TR | 23-kHz ultrasonic transducer |

# High-performance alarm

The high-performance home security alarm control-box circuit shown in Fig. 7-26 provides excellent performance and reliability at low cost. The circuit provides some advanced alarm features as well. The high-performance alarm circuit allows for

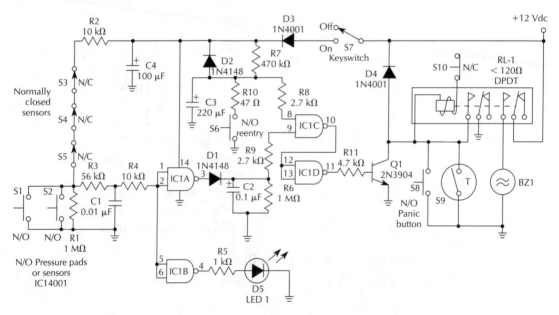

*Fig. 7-26  High-performance alarm system.*

many different types of sensors to be used within this system. Inputs are provided for normally open switches or sensors such as tape switch mats, as well as normally closed-loop-type sensors such as window foil, magnetic reed switches, and people sensors, including passive infrared detectors. In addition, heat sensors or thermostats can also be used, and they are given instant protection status, day or night. The high-performance alarm circuit draws only a few microamperes under idle conditions.

Operation of the high-performance alarm circuit begins with S7 closed and all sensors inactive. LED1, the relay, and the buzzer are off. Low-pass filters R3/C1 and RG/C2 suppress any spikes or transients introduced to or near the alarm circuit, to avoid false alarms. If any of the sensors or switches become activated, the outputs of IC1A and IC1B go high, thus turning on LED1, the buzzer, and relay. The relay is activated through Q1 via IC1C and IC1D. As the relay turns on, it self-latches via the left set of relay contacts. The self-latching relay contacts are permanently wired to 12 V. The relay can be activated immediately by the panic button or by the thermostatic switch S9. If the keyswitch is initially closed, or if the push button S6 is pressed and released, the R7/C3 network disables Q1 for about 100 seconds. At the end of this period, the circuit returns to normal operation. This feature is of great value, because when the system is first turned on via S7, it keeps LED1 off.

The reentry switch S6 allows you to enter/exit without setting off the alarm. You should install a keyswitch at your most accessed entry door. Switches S3, S4, and S5 can be any type of normally closed sensors or switches. You could wire any number of additional sensors in series with these three switches. Switches S1 and S2 are normally open switches or sensors such as tape switch mats. You could wire any number of normally open sensors in parallel across these switches as well. The normally open panic switch S8 can be paralleled with other panic switches distributed throughout your home or office. A triple-pole relay could be substituted for the relay shown if you need extra contacts to activate a telephone dialer. If you wish to have a silent alarm, wire the contacts used for the buzzer instead to a dialer. The alarm control circuit should include a 12-V ni-cad battery charged via line current to ensure reliable operatue during a storm or power outage.

Multiple loop circuits consisting of additional gates such as IC1A and IC1B can be created and combined at IC1C pin 9. In this way you can create separate loops, one for perimeter protection and a different loop for space protection. Each loop then would have its own LED indicator lamp to aid in troubleshooting system loop problems. The high-performance alarm circuit is a starting point in creating your own custom alarm system.

## High-performance alarm system parts list

| Quantity | Part | Description |
|---|---|---|
| 2 | R1, R6 | 1-M$\Omega$, ¼-W resistor |
| 2 | R2, R4 | 10-k$\Omega$, ¼-W resistor |
| 1 | R3 | 56-k$\Omega$, ¼-W resistor |
| 1 | R5 | 1-k$\Omega$, ¼-W resistor |
| 1 | R7 | 470-k$\Omega$, ¼-W resistor |
| 1 | R8, R9 | 2.7-k$\Omega$, ¼-W resistor |
| 1 | R10 | 47-$\Omega$, ¼-W resistor |
| 1 | R11 | 4.7-k$\Omega$, ¼-W resistor |
| 1 | C1 | 0.01-$\mu$F, 25-V capacitor (disk) |
| 1 | C2 | 0.1-$\mu$F, 25-V capacitor |
| 1 | C3 | 220-$\mu$F, 25-V electrolytic capacitor |
| 1 | C4 | 100-$\mu$F, 25-V electrolytic capacitor |
| 2 | D1, D2 | 1N4148 silicon diode |
| 2 | D3, D4 | 1N4001 silicon diode |
| 1 | D5 | Red LED |
| 1 | Q1 | 2N3904 transistor |
| 1 | IC1 | CD4001 CMOS IC |
| 1 | RL-1 | 120-$\Omega$, 12-V, DPDT relay |
| 1 | BUZ | Piezo buzzer |
| 2 | S1, S2 | Normally open sensors or switches |
| 3 | S3, S4, S5 | Normally closed sensors or switches |
| 1 | S6 | Normally open reentry push-button switch |
| 1 | S7 | Keyswitch |
| 1 | S8 | Normally open panic button |
| 1 | S9 | Normally open thermal sensor |

# 8
# Unique high-tech security projects

THE FUN AND CHALLENGE OF CONSTRUCTING NEW AND NOVEL circuits from the ground up is often the only answer to your new ideas, techniques, or projects. If you are skilled in electronic project construction, you will find the complete schematics, parts lists, and printed-circuit layouts of tremendous value when deciding to build one of the high-tech security projects.

I hope this cookbook has either informed or inspired you and that it will serve your current and future sensing and security needs.

## Piezo vibration alarm

A sensitive, low-cost piezo vibration sensor could solve many of your sensing and alarm problems. The piezo vibration alarm can protect your home, office, and automobile. You can construct the piezo vibration sensor to serve as a compact "space" or interior area protection device, or as a "spot" type sensor to protect specific items such as antiques or expensive electronics. A number of these small vibration sensors could be wired together and connected to a central alarm control panel to form a complete alarm system. You could also configure the piezo vibration sensor as a self-contained portable travel alarm by substituting an SCR in place of the relay-driving transistor Q1, as shown in Fig. 8-1.

The heart of the piezo vibrator sensor is a small, low-cost Kynar piezo film sensor (DT1-028K) from Penwalt Corporation. The sensor is a thin-film Kynar plastic bonded to a thin strip of spring steel. The sensitivity of the sensor can be further increased by mechanically attaching a very small weight to the "free" end of the metal sensor strip. The external stress flexes the metal strip

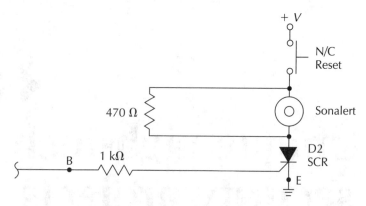

**Fig. 8-1** *Piezo vibration alarm.*

with a proportional output, and the resultant signal is fed into an op amp to drive a relay. The piezo sensor usually protects the up/down or Z axis; however, by placing the sensor on its side, the sensor could protect or monitor the horizontal axis as well.

Operation of the piezo vibration sensor begins with the piezo film sensor, which is coupled to a Texas Instruments TLC271 op amp configured as a nonretriggerable monostable multivibrator. The output of the op amp on pin 6 drives an npn transistor Q1, which drives a low-current relay, as shown in Fig. 8-2. The relay can be connected to an alarm control panel. When the signal

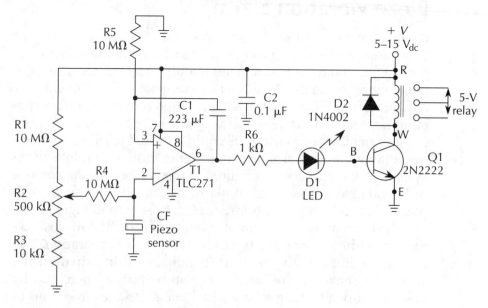

**Fig. 8-2** *Multiple sensor hookup.*

from the film sensor exceeds the predetermined level set by R2, a 500-kΩ potentiometer, the op amp triggers, activating the series LED and the low-current relay (Radio Shack 275-243). The series LED indicator can be quite useful in the setup and testing of the vibration alarm. The piezo vibration alarm sensor can be custom configured in a number of different ways. First, by changing the values of R1, R2, and R3, the gain values for the circuit can be increased or decreased. Adjust the alarm "ON" time by changing the values of C1 and R5. Change the low-frequency cutoff by altering the value of R4.

The vibration sensor circuit has a 2-second recovery period before the next trigger event can be recorded. The peak output voltage of the sensor is directly proportional to the peak acceleration. A minimum deflection-factor sensitivity of 8 mG and a maximum sensitivity of 680 mG can be expected when using the component values shown and a 9-V power source. The piezo alarm circuit can be powered from a 5- to 15-V source.

The piezo alarm can drive a relay, as shown in Fig. 8-2, or it can be used as a portable travel alarm by substituting a small SCR in place of the relay driver Q1, as shown in Fig. 8-1. A sonalert is driven by the SCR in series with a normally closed reset button. The circuit can be powered by a 9-V transistor-radio battery, because power consumption is very low.

When constructing a multiple-sensor system (see Fig. 8-3), the sensor enclosure can be quite small. A box measuring 2½ × 1½ inches could house a complete sensor unit. Power, signal, and ground wires are connected in parallel between each of the sensor units wired back to the central alarm panel. If you need to display specific location information, all of the signal wires

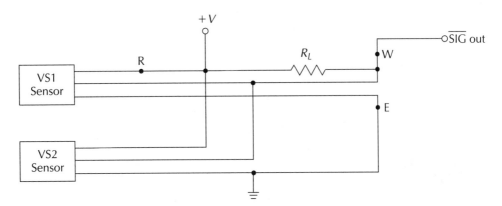

**Fig. 8-3** *Piezo-vibration-alarm latching circuit.*

would have to be returned separately back to the control panel. A slightly larger box would be required if you plan on using the piezo alarm as a portable self-contained latching travel alarm. The portable alarm could be placed in a motel or apartment when you are traveling. Simply place the sensor unit near an outside-facing wall or door area and adjust the sensitivity control.

Often the sensitivity of vibration sensors is carelessly set to maximum without any regard to the local surroundings or noise conditions. In most applications, set the sensitivity to maximum and decrease it by 10 to 15 percent. If you are using the piezo vibration as a "spot" or specific item protection device, set the gain to one half. If you mount vibration sensors on windows, be very careful in setting up the sensors, because windows flex during storms and windy conditions. Take your time and adjust each sensor in its location to minimize the chances of false alarms. Don't annoy your neighbors or the police with false alarms! The vibration alarm needs extra care in setup and testing to ensure reliable and troublefree operation. The piezo vibration sensor is low cost, but effective when used sensibly, and it can solve a number of your alarm needs.

**Piezo vibration sensor parts list**

| Quantity | Part | Description |
|---|---|---|
| 1 | R1 | 10-M$\Omega$, ¼-W, 5% resistor |
| 1 | R2 | 500-k$\Omega$ potentiometer (linear taper) |
| 1 | R3 | 10-k$\Omega$, ¼-W, 5% resistor |
| 1 | R4 | 10-M$\Omega$, ¼-W, 5% resistor |
| 1 | R5 | 10-M$\Omega$, ¼-W, 5% resistor |
| 1 | R6 | 1-k$\Omega$, ¼-W, 5% resistor |
| 1 | C1 | 223-pF capacitor |
| 1 | C2 | 0.1-µF capacitor |
| 1 | CF | DTK028k Penwalt piezo film sensor |
| 1 | D1 | LED |
| 1 | D2 | 1N4004 diode |
| 1 | Q1 | 2N2222 transistor |
| 1 | IC1 | TLC271 IC (Texas Instruments) |
| 1 | RLY | 5-V SPDT relay (Radio Shack 275-241) |

# Camp alarm

The camp alarm is a novel compact perimeter alarm system designed for campers, hunters, and outdoor persons. The camp alarm provides you with an advance warning of approaching humans and animals, and will help you to obtain a good night's sleep in the great outdoors. The camp alarm is a dual-input capacitive alarm that uses two low-power versions of the ubiquitous 555 chip to sense movement. The camp alarm has two input circuits with two different charge accumulator capacitor inputs, C1 and C2. The different input capacitor values allow one input to accumulate a charge ahead of the other channel, as a type of discriminator. Both channels must be activated at the same time to activate the alarm buzzer (see Fig. 8-4).

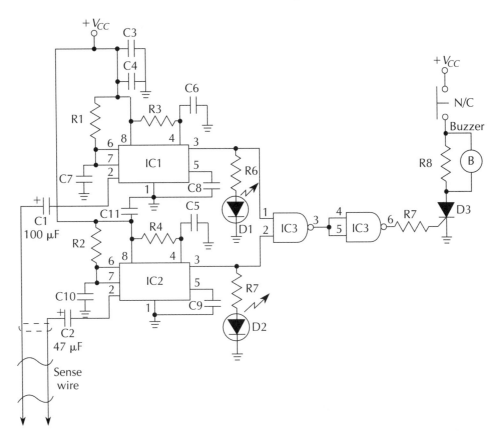

**Fig. 8-4** *Camp alarm.*

The human body represents a capacitance of about 300 pf, referenced to ground. The body acts as an antenna that is loosely coupled to a ground path. By approaching the antenna or sense wire, the body capacitance effectively lowers the input impedance on the input pin 2 of IC1 and IC2 to below the threshold value. The

R1/C7 and the R2/C10 combinations provide the output pulse time values for each of the detectors. The R3/C6 and R4/C5 pairs provide power-on reset (POS), which prevents triggering of the detectors upon power-up. The two LEDs show how the two inputs react to triggering. They can be eliminated, if desired. Once both IC1 and IC2 trigger simultaneously, the signals are combined by IC3, a 7400 series NAND gate, and, in turn, an SCR is fired, sounding the alarm buzzer. Once triggered, the SCR will remain in the "on" condition until switch S2 resets the alarm. Switch S2 is a normally closed push button. Whether you select an electronic buzzer or Sonalert for the alarm sounder determines if you need resistor R8. If you select a Sonalert sounder, R8 is required, because the Sonalert might not draw enough current to keep the SCR latched. If you wish to build a nonlatching or momentary-type alarm that could be used as an annunciator alarm, substitute a 2N3904 or equivalent npn transistor in place of the SCR. The transistor drives the sounder, and you can eliminate switch S2.

Construction of the camp-alarm project is quite easy. A G-10 glass-epoxy, 1-oz circuit board was used for the prototype unit. The circuit was placed in a small plastic box measuring 1⅞ × 2¼ × 3 inches. A 9-V alkaline transistor-radio battery powered the camp-alarm circuit. Place the antenna or sense wire jack far enough away from the on-off switch to prevent false triggering of the system upon power-up.

The sense wire or antenna is comprised of thin-gauge speaker wire such as Belden 8782, an extremely plyable wire. The antenna/sense wire is looped around four dowels, as shown in Fig. 8-5. The 12- to 15-inch-long by ¼-inch-diameter dowels are tapered at one end to ease placement into the ground. Wrap four to five turns of sense wire around each dowel to form a square perimeter around your campsite. The four dowels provided a complete self-contained alarm system that could be used anywhere, including the beach where there are no trees. The scheme is to form a complete square perimeter around your campsite, thus providing you with ample time before an animal or person enters your sleeping space. This extra time permits you to make a quick decision in order to take necessary action.

Experiment with the antenna/sense wire placement and height and with the values of C1 and C2 to suit your particular needs. Quick, abrupt movements near the sense wire seem to trigger the system most reliably. The camp alarm can also be used for other applications, including annunciators to detect persons entering a room or store front. The camp alarm could

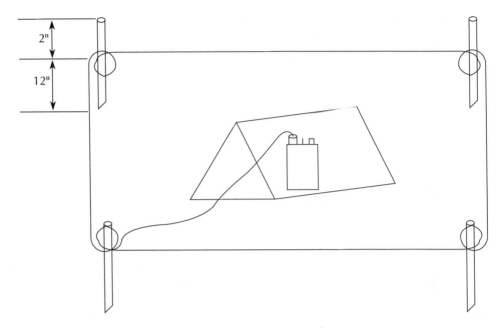

*Fig. 8-5*  *Camp alarm setup diagram.*

also be used to activate a point-of-purchase display in a retail store, or to activate a motor, lamp, or display in a museum exhibit. When configuring the camp alarm to serve as an annunciator, substitute an npn transistor as mentioned earlier, and increase the values of capacitors C7 and C10, if necessary, to lengthen the "on" time of the timers. You might find other uses for the camp alarm.

### Camp-alarm parts list

| Quantity | Part | Description |
|---|---|---|
| 2 | IC1, IC2 | 1555 low-power timer |
| 1 | IC3 | SN7400 NAND gate |
| 2 | R1, R2 | 100-k$\Omega$, ¼-W resistor |
| 2 | R3, R4 | 24-k$\Omega$, ¼-W resistor |
| 2 | R5, R6 | 330-$\Omega$, ¼-W resistor |
| 1 | R7 | 1-k$\Omega$, ¼-W resistor |
| 1 | R8 | 470-$\Omega$, ¼-W resistor |
| 2 | C1, C3 | 100-$\mu$F capacitor, 25 WV$_{dc}$ |
| 1 | C2 | 30-$\mu$F capacitor, 25 WV$_{dc}$ |
| 1 | C4 | 0.1-$\mu$F capacitor, 25 WV$_{dc}$ |
| 2 | C5, C6 | 1-$\mu$F capacitor, 25 WV$_{dc}$ |
| 2 | C7, C10 | 2.2-$\mu$F capacitor, 25 WV$_{dc}$ |

| Quantity | Part | Description |
|---|---|---|
| 2 | C8, C9 | 0.05-µF capacitor, 25 $WV_{dc}$ |
| 2 | D1, D2 | Red LEDs |
| 1 | D3 | NTE5404 SCR |
| 1 | B | Electronic buzzer or Sonalert |
| 1 | S1 | SPST on-off power switch |
| 1 | S2 | Normally Closed push-button switch |
| 1 | MISC | Plastic box, battery clip, 9-V battery, sense wire, RCA jack |

## Pyroelectric sensor

The pyroelectric detector is an excellent starting point for constructing your own home-alarm system. The infrared body-heat detector has proven itself in the field and has become the alarm industry's preferred sensor, with good sensitivity, little energy consumption, and great reliability. The sensitive pyroelectric sensor (PES) can be used to sense humans and animals up to 50 feet away. The pyroelectric sensor is generally used with a low-cost Fresnel lens ahead of the sensing element to provide the desired coverage. The thin lens comes in a variety of types that can be used in a long range configuration, providing a narrow beam coverage up to 50 feet. At the other end of the coverage spectrum, a lens is available that provides a wide pattern of coverage out to 12 feet (see Fig. 8-6). Special lenses are also available to avoid pets in the protected area. Pyroelectric sensors can also be used without a lens to protect specific valuables or as a touch switch to activate a store display. The PES can also be used for lighting control, flame sensors, door controls, robotics, and animal sensors.

The PES can be used indoors as well as outdoors. When used outdoors, the sensor and its electronics must be housed in a watertight box. The PES system is shown in Fig. 8-7. It uses a low-cost, dual-element, parallel-opposed sensor that costs only about $5. Signals from a radiation source falling on both active sense elements will be canceled; however, when a source of radiation passes from one element to the other, two separate outputs are produced. The Eltec 5192 IR pyroelectric sensor is a dual-element lithium-tantalate sensor housed in a TO-5 package, with a small IR filter placed in front of the sensing elements. Pyroelectric detectors respond more rapidly than most other IR sensors. They respond to changes in IR intensity rather than just the presence of an infrared source, allowing them to be used in many applications.

A number of unsymmetrical films or crystals such as lithium

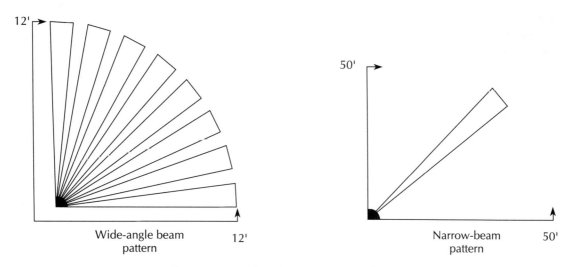

**Fig. 8-6** *Pyroelectric sensor pattern/range.*

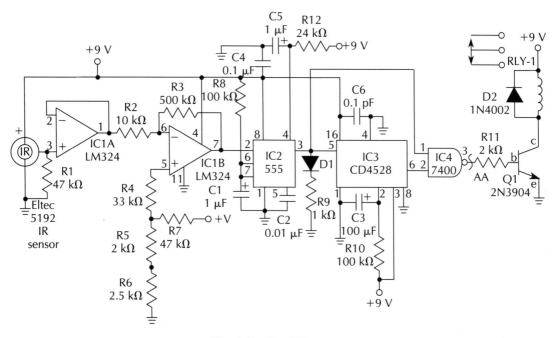

**Fig. 8-7** *The PES system.*

tantalate demonstrates the pyroelectric effect. A thin wafer or film is coated on both sides, forming electrodes and thus forming a type of capacitor with a capacitance of about 30 pf. The pyroelectric material exhibits an internal electric field collected between the two electrodes. To monitor the resultant charge requires a high-impedance amplifier. Therefore, most sensors incorporate a high-value load resistor, providing an impedance converter as well as a field-effect transistor amplifier packaged along with the sensing elements.

The circuit description begins with a 47-kΩ source resistor that sets the drain current. It is placed across the signal output and ground. The signal is fed to a voltage follower, which minimizes loading to the sensor. The voltage follower is formed by IC1A. The output of IC1A is coupled via a 10-kΩ resistor to IC1B, the second half of an LM324 op amp. A 500-kΩ resistor between pins 6 and 7 sets the overall gain of the amplifier.

Pin 5 of IC1B is connected to a threshold of set-point control formed by voltage divider R5, R6, and R7. The output of IC1B is sent to pin 2 of the 555 timer IC. The timer "on time" is controlled by the R8/C1 combination, as shown. The power-on reset (POR) is formed by R12/C5, provided on pin 4 of the 555 timer. The output signal on pin 3 couples to pin 5 of IC3, a dual retriggerable monostable CD4528. The CD4528 creates a time window and forms a discriminator. The output from the 555 is also coupled to a 7400 NAND gate along with the output from the CD4528. Thus, by adjusting the time window via C3/R10, only a certain number of pulses from the 555 is allowed through the window set by the CD4528.

The LED D1 helps to set up the sensor and aids to adjust the threshold control R5. The output of the 7400 on pin 3 (also point AA) is fed to a 2-kΩ resistor driving a 2N3904 npn transistor, which drives a low-current relay. The output relay can trigger an alarm control panel. Another approach would be to use an SCR in place of Q1 and to drive a Sonalert sounder instead of the relay, as shown in Fig. 8-8. Once triggered, the Sonalert sounds until reset by S1, a normally closed push-button switch. You could now use the PES as a self-contained travel alarm.

Construction of the PES was completed using a G-10 glass-epoxy circuit board. The 5192 sensor should be mounted as close to IC1A as possible. Mount the completed circuit board in a small metal box to eliminate stray rf energy and thermal noise, including air currents, from affecting the sensors operation. The pyroelectric sensor is sensitive to both rf and air current. Therefore, the sensor should be in an airtight metal box.

Mount the Fresnel lens in front of the sensor, as shown in Fig. 8-9. Bend the lens in an arc, with the sensor placed 1.2 inches behind the focal point of the lens. Mounting the lens is somewhat problematic, because the lens is bent in an arc and most boxes

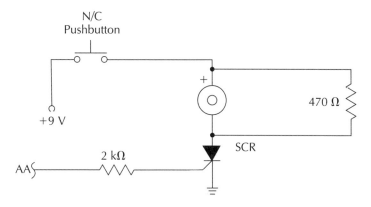

**Fig. 8-8**  *Pyroelectric sensor latch circuit.*

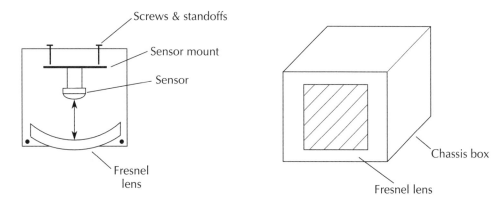

**Fig. 8-9**  *Pyroelectric sensor and lens mounting.*

are rectangular. Remember that air currents must be kept from the detector. The lens is placed over a cutout on the chassis box and bent in an arc.

Another approach to this problem would be to mount the sensor over a cutout on the side of a tuna can (see Fig. 8-10). Mount the sensor behind the lens at the 1.2-inch focal distance. A shielded wire would then connect the sensor to the electronic circuit board placed in a box mounted under the tuna can. You might discover another mounting method better suited to your application.

The pyroelectric sensor circuit can be powered from an 8- to 12-$V_{dc}$ source, such as a small dc wall-cube power supply or from an alarm panel. If the sensor is used as a self-contained portable alarm, a 9-V transistor-radio battery could be used. A number of these pyroelectric sensors could be wired in parallel to form a complete alarm system.

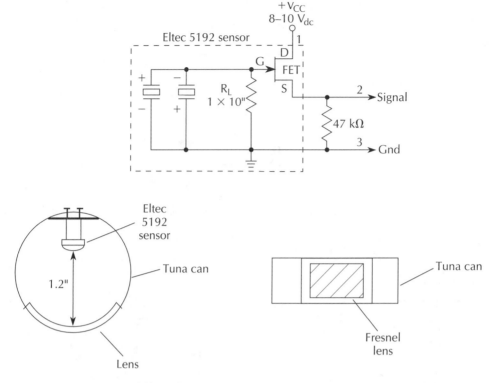

**Fig. 8-10** *Pyroelectric sensor tuna-can mount.*

Operation of the PES is quite simple. First, apply power. After a few seconds, place your hand in front of the sensor and adjust the threshold control R5 until the LED comes on when your hand passes in front of the sensor. The LED should go out after a few seconds. If it doesn't, back down R5 until it does. Set the discriminator to allow about two pulses through in a 10- to 15-second time period. This is controlled by R10, which can be adjusted if necessary. The IR pyroelectric sensor is now ready to stand guard and protect your home or office. The body heat or pyroelectric detector has proven itself in the field and makes a good choice for a troublefree sensor.

## Pyroelectric sensor parts list

| Quantity | Part | Description |
|---|---|---|
| 1 | IR | 5192 Eltec pyroelectric sensor |
| 1 | R1 | 47-k$\Omega$, ¼-W, 5% resistor |
| 1 | R2 | 10-k$\Omega$ resistor |
| 1 | R3 | 500-k$\Omega$ resistor |

| | | |
|---|---|---|
| 1 | R4 | 33-kΩ resistor |
| 1 | R5 | 2-kΩ potentiometer, 10 turns |
| 1 | R6 | 2.5-kΩ resistor |
| 1 | R7 | 47-kΩ resistor |
| 1 | R8 | 100-kΩ resistor |
| 1 | R9 | 1-kΩ resistor |
| 1 | R10 | 100-kΩ resistor |
| 1 | R11 | 2-kΩ resistor |
| 1 | R12 | 24-kΩ resistor |
| 1 | C1 | 4.7-µF, 25-V electrolytic capacitor |
| 1 | C2 | 1-µF, 25-V capacitor |
| 1 | C3 | 100-µF, 25-V electrolytic capacitor |
| 1 | C4 | 0.1-µF capacitor |
| 1 | IC1 | LM324 IC |
| 1 | IC2 | LM555 IC |
| 1 | IC3 | CD4528 IC |
| 1 | IC4 | 7400 IC |
| 1 | RLY1 | 5-V relay (Radio Shack 275-241) |
| 1 | Q1 | 2N3904 transistor |
| 1 | D1 | LED |
| 1 | D2 | 1N4001 diode |

# Chimney alarm

The chimney alarm is a novel fire-alarm system that will set your mind at ease and perhaps save you insurance money as well. The chimney alarm is a multipurpose fire alarm that will sound a local alarm, call the fire department, and simultaneously dump dry chemical powder down the chimney flue pipe to begin extinguishing the fire before it can get out of control. The chimney-alarm system also allows additional switches or sensors to be connected, such as rate-of-rise thermostats or smoke detectors, which you could place throughout your home.

The chimney alarm is quite straightforward and is based upon a normally closed or "supervised" thermo-disc snap-action thermostat sensor switch, which is designed to open its contacts at 575°F. The snap-action sensor switch was chosen because it is extremely reliable and resettable over long periods of time. In addition, the sensor is rugged and easy to install. The use of the thermo-disc switch decreases the overall parts count and increases the reliability of the system.

The circuit description begins with a 7400 (see Fig. 8-11) NAND gate, which permits the "supervised" input from activat-

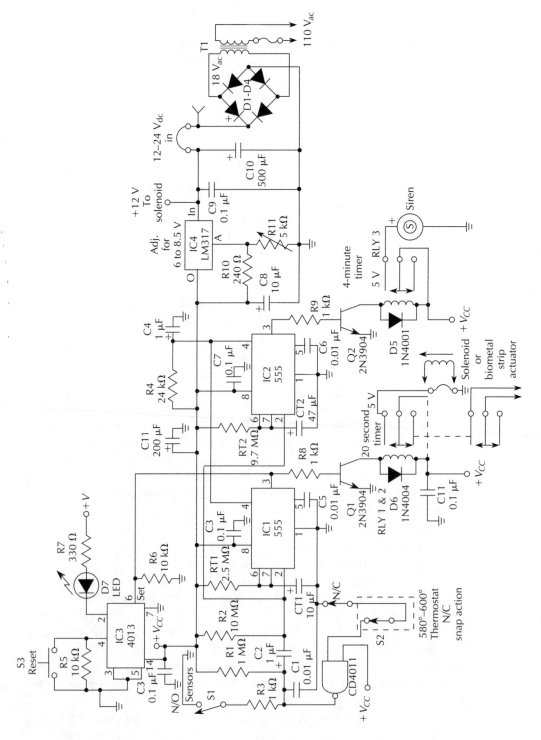

Fig. 8-11 Chimney alarm.

ing the 555 timer. Two 555 timers are configured as manually triggered monostable multivibrators. Prior to activating or triggering, C1 is charged to the supply voltage through R1. Triggering the sensors discharges C1 rapidly through R3, creating a short negative spike. Any spurious spikes are removed by the integration action of R1 and C1. The resultant clean negative pulse is passed through C2 to trigger the 555 timers. Once fired, both 555 timers, which are connected together at pin 2, generate an output pulse determined by the formula $T = 1.1\, R_T/C_T$. The input trigger circuit triggers only once per activation. Therefore, regardless of how long the sensors remain open or closed, a finite output is produced. Timer A is set up to pull in RLY1 via a 2N3904 npn transistor for a period of 20 to 30 seconds. This time period is enough to trigger a telephone dialer and to activate the chemical dumper that drops the powder into the chimney. Timer B, also a 555 timer, activates a 2N3904 that pulls in RLY2 for a period of 4 to 5 minutes, depending on the Rt2/Ct2 combination, to power a siren or Sonalert. Activating the timer also triggers a memory circuit, a CD4013 flip-flop, which indicates if the system has been triggered.

The memory-indicator LED remains on until S3, a normally open reset button, is reset.

The thermo-disc sensor can be easily mounted on a metal chimney flue pipe by screwing the sensor's flanges directly to the flue pipe about 3 to 5 feet from the fire box. The sensor could also be mounted via an adjustable automotive hose clamp. A brick or tile chimney presents another type of mounting problem. The sensor could be mounted inside the firebrick at a 3- to 5-foot distance up the flue. Another approach would be to drill a hole into the firebrick and use a heat-transfer pipe to transfer the heat to the sensor, which is mounted on the outside face of the fireplace tower. Figure 8-12 presents a conceptual mechanical design for the chemical powder dumper, but final design details are left to the builder. The diagram gives a possible design based on a biometal actuator using a bell housing and spring arrangement. A solenoid could also be used in place of the biometal wire. The biometal wire or Nitinol wire is available from Toki or Monotronics, listed in the Appendix.

By applying a voltage for a short period of time, the metal wire expands or contracts. A 2- to 3-inch piece of biometal requires about 400 mA of current to contract the wire. An activation period of 15 seconds is needed to expel the dry chemical powder.

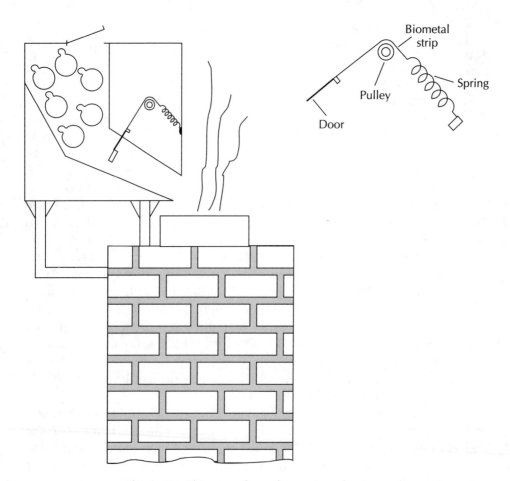

**Fig. 8-12** *Chimney alarm dumper mechanism.*

Dry chemical powder has long been used by fire departments to extinguish fires. It is readily available at a relatively low cost. Contact your local fire department on availability and cost. Small plastic bags of dry powder could be housed in the dumper mechanism until the alarm is triggered. The metal dampers should probably be insulated so as to not destroy or overheat the bags, which hold the powder.

Another approach to the chimney-alarm system would be to use a solenoid to "pull a pin" from a fire extinguisher, which would force the tank's contents into the chimney from either above via the roof or directly into the firebox or chimney from inside your home. Either scheme to extinguish fires depends upon the user to refill the extinguisher or dry-powder bags after each event. Construction of the chimney alarm begins by using a 3- by 6-inch glass-epoxy circuit board.

All components except the siren and power transformer can be mounted on the circuit board. The on-off switch, reset button, and LED are mounted on the front cover of a metal box measuring 5 × 8 inches. The siren or Sonalert can also be mounted on the front cover. A screw-terminal barrier strip mounted on the rear of the chassis accepts the sensor switches and output connections from the relay outputs.

A power supply for the chimney alarm consists of a 12–18-$V_{ac}$ transformer connected to the bridge rectifier, as shown in Fig. 8-11. The filtered dc output from the bridge is applied to an LM317 regulator, adjusted for 8–10 V by the 5-k$\Omega$ potentiometer. An alternative supply could be a 6–9-$V_{dc}$ wall cube. However, pay attention to the current requirements of the output device. A Sonalert requires much less current than a siren, for example. As mentioned, the biometal activator or solenoid requires about 400–500 mA of current, so your power supply should provide at least 1 A.

The dry chemical dumping mechanism, shown in Fig. 8-12, can be used to open the door of a metal fabricated dumper box. A moderately tight spring is used against the force of the extended Nitinol wire in the bell crank assembly.

Operating the chimney alarm is simple, once all the sensors have been mounted and the outputs connected. Simply apply power. If the memory LED is on, press S3 to reset the lamp to test the circuit, short the normally open contacts, or open the normally closed loop to ensure the system is working OK. Test the system adequately before connecting a telephone dialer to ensure any false alarms from the system will not call the fire department. The chimney alarm is now ready to protect your home and its contents.

## Chimney alarm parts list

| Quantity | Part | Description |
|---|---|---|
| 1 | R1 | 1-M$\Omega$ ¼-W, 5% resistor |
| 1 | R2 | 10–M$\Omega$, ¼-W, 5% resistor |
| 3 | R3, R8, R9 | 1-k$\Omega$, ¼-W, 5% resistor |
| 1 | R4 | 24-k$\Omega$, ¼-W, 5% resistor |
| 2 | R5, R6 | 10-k$\Omega$, ¼-W, 5% resistor |
| 1 | R7 | 300-$\Omega$, ¼-W, 5% resistor |
| 1 | R10 | 240-$\Omega$, ¼-W, 5% resistor |
| 1 | R11 | 5-k$\Omega$ potentiometer (trim) |
| 1 | Rt1 | 2.5-M$\Omega$ (see text) |

| Quantity | Part | Description |
|---|---|---|
| 1 | Rt2 | 4.7-MΩ resistor (see text) |
| 3 | C1, C5, C6 | 0.01-µF, 25-$V_{dc}$ capacitor |
| 2 | C2, C4 | 1-µF electrolytic, 25 $V_{dc}$ |
| 3 | C3, C7, C9 | 0.1-µF 25-$V_{dc}$ capacitor |
| 1 | C8 | 10-µF electrolytic 25 $V_{dc}$ |
| 1 | C10 | 500–1000-µF electrolytic, 25 $V_{dc}$ |
| 1 | C11 | 200-µF electrolytic, 25 $V_{dc}$ |
| 1 | Ct1 | 10-µF electrolytic, 25 $V_{dc}$ |
| 1 | Ct2 | 47-µF electrolytic, 25 $V_{dc}$ |
| 4 | D1–D4 | 1N4001 diode |
| 2 | D5, D6 | 1N4004 diode |
| 1 | D7 | Red LED |
| 3 | RLY1–RLY3 | 5-V relay (Radio Shack 275-241) |
| 2 | IC1, IC2 | 555 timer |
| 1 | IC3 | CD4013 flip-flop |
| 1 | IC4 | LM317 regulator |
| 1 | IC5 | CD4011 NAND gate |
| 2 | Q1, Q2 | 2N3904 transistor |
| 1 | S | Sonalert, 5–9 $V_{dc}$ |
| 1 | T1 | 12–14-V transformer |

# DTMF alarm system

The remote identification alarm is a novel 15-station DTMF (dual-tone multifrequency) ID alarm system that can easily be expanded to meet most alarm needs. The remove ID alarm is quite flexible, and it can be used over both hardwire or rf radio links to identify specific alarm locations upon triggering. Figure 8-13 illustrates an alarm trigger sending module, consisting of a trigger circuit and a dual-tone oscillator. Both normally open and normally closed type switches or sensors can be connected to the input of the sending module. The input of the trigger section uses a 555 timer configured as a monostable multivibrator. A CD4011 NAND gate is set up to accept various sensors as shown. The capacitor (C2) on the input pin 2 is initially charged and, as the NAND gate is activated, a short negative pulse is generated, turning the 555 on for a time period determined by the R4/C4 combination. This time period is set for about 15 seconds. The output of the trigger section on pin 3 of IC1 is coupled to an NTE159 transistor, which turns on the dual-tone oscillators. Each sending module is set up to generate a dual-tone frequency pair, as shown in Fig. 8-13. The first tone oscillator must produce a 697 Hz signal in IC2A, and IC2B must be

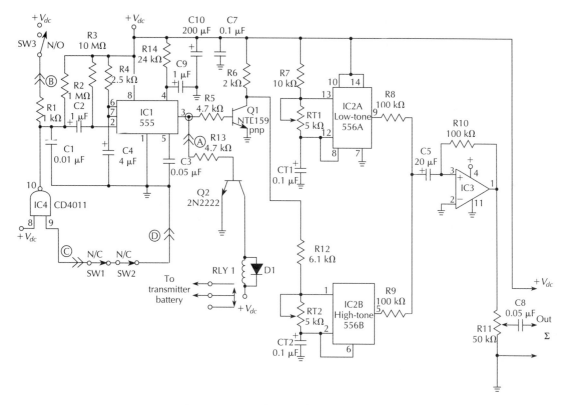

**Fig. 8-13** Dual tone alarm transmitter.

adjusted to produce a 1209-Hz signal to activate a valid output on the tone decoder corresponding to digit number 1 or location 1 on the display panel.

The DTMF ID alarm can be used over a three-conductor hardwire system, with each of the sending modules wired in parallel with each of the other sending units, as shown in Fig. 8-14. An rf link system could be used instead of the hardwire hookup. The DTMF audio output from the sending module is coupled to the audio input of a small rf transmitter in the 49-MHz band, in the 108-MHz broadcast band, or on the VHF itinerant communication frequencies of 154.625 MHz or 154.570 MHz. Various transmitter modules are readily available and sources are listed in the Appendix. The audio level control on the sending module is adjusted to modulate the rf transmitter, but take care to ensure that the transmitter will not be overmodulated. Many of the small FM broadcast-band transmitters can be used and easily modified to operate on other nearby frequencies. In this configuration, a relay is added to the sending module at point AA or pin

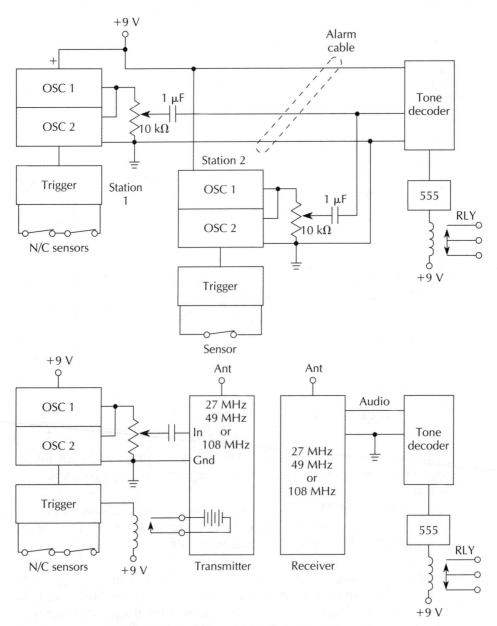

**Fig. 8-14** *DTMF multisensor hookup diagram.*

3 of IC1. A low-cost npn transistor would be used to drive a low-current relay to apply power to the transmitter for a 15-second duration, to allow the receive decoder to decode the DTMF audio tones. Any receiver on the same frequency as the transmitter may be connected to the input of the decoder, as shown in Fig. 8-15.

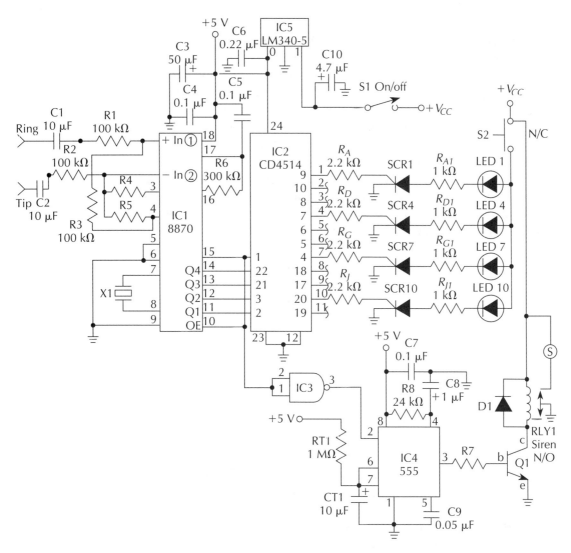

**Fig. 8-15** *Dual tone receiver/decoder/display unit.*

The receive decoder consists of a low-cost Teltone 8870 DTMF decoder IC, coupled to a CMOS CD4514, a 4- to 16-line decoder providing 15 outputs or alarm locations. Each of the 15 outputs drives a small SCR that energizes the LED display. The steering-logic output on pin 15 of IC1 goes high as a valid tone pair is received. This signal activates IC4 at the 555 timer via an inverter. The 555 controls an output device, such as a siren bell or automatic telephone dialer. The "on timer" of the output device is controlled by the Rt1/Ct1 combination, and it is adjusted for a 4-minute time period.

The DTMF location alarm is constructed using G-10 glass-epoxy circuit boards. The tone modules are built on a 2 × 3-inch circuit board. The sending-module circuit board also has provisions for the transmitter control relay, so you can populate this section if you anticipate using rf radio links. The normally closed switches or sensors are all wired in series between point C and D on the circuit diagram. The normally open switches are connected from point B to the power bus or supply. The point labeled E is the tone output signal connected either to a transmitter or hardwire cable as discussed earlier. The most difficult part in constructing the ID DTMF location alarm is adjusting the tone oscillators in the sending modules. Each tone pair must be adjusted quite closely to the DTMF frequencies listed in Table 8-1.

Table 8-1 Tone selector guide.

| Low Tones | 1209 | High tones 1336 | 1477 | Hz 1633 |
|---|---|---|---|---|
| 697 | 1 | 2 | 3 | A |
| 770 | 4 | 5 | 6 | B |
| 852 | 7 | 8 | 9 | C |
| 941 | * | 0 | # | D |

Power for the tone modules can be obtained in several ways. If the modules are going to be used in a hardwired system, the three-wire cable to each module could supply power from either the control alarm control panel or from a separate 12- or 6-V power supply. If the sending modules are to be used with rf transmitters at each location, batteries could be used for each trigger module and transmitter combination, or a dc wall-cube supply could charge the batteries or directly power the circuits. Powering the circuits is left to the reader and depends on the final application.

The current rating of the power supply would depend upon the power output of the transmitter used. A 100-mW transmitter draws far less power than a 0.5-W transmitter.

The DTMF decoder/display unit is rather straightforward in construction. The decoder requires no setup or alignment procedure. Care must be taken, however, to prevent overmodulation and/or distortion from reaching the decoder. Carefully check to make sure that the audio level on the receiver is not adjusted too high. The decoder/display unit is constructed on a 6 by 6-inch glass-epoxy circuit board that holds both the decoders and display. You can install all of the SCRs and respective LED display lamps. If you wish to construct a starter system, you could populate the extra station components as needed. The output relay RLY1 is also mounted on the circuit board, and it can be connected directly to a Sonalert or siren. RLY1 could also be wired to a power relay if you anticipate a high-current motor siren. If you plan on using an automatic telephone dialer in your system, adjust the time period of Ct1/Rt1 to a shortened time interval, to ensure the dialer will not be triggered more than once.

You can power the receiver/decoder by a few different methods. You could purchase a surplus 6- to 9-V or adjustable power supply or you could build a supply. Another approach would be to power the decoder unit via a 6-V nicad battery, trickle-charged by a wall-cube power supply. An LM340-5 regulator on the decoder/display unit provides 5 V to the DTMF-decoder chip, a voltage-sensitive 8870. If you are going to power all the sending modules and the decoder/display unit along with a high-powered siren in a hardwire cable system, anticipate the power requirements of the entire system when choosing your power supply. An rf system would require much smaller power supplies at each location, and the power requirement for the decoder unit would be considerably less.

Operating the DTMF ID location alarm is simple and straightforward. Having installed all the tone sending modules in their respective locations, apply power to the modules and to the decoder/display unit. Press the reset buttons on the decoder unit to ensure all display lamps are off. Go over to one of the sending modules and trigger it by connecting point B to the + lead or power bus. The sending modules should now produce a DTMF pair for 15 seconds, and the transmitter relay should also be activated, if installed.

The trigger "ON" time can be increased while you are adjusting or troubleshooting the system. The trigger time is adjusted by R4/C4 on the transmitter module. No special setup is required for the decoder unit. Once power is applied to the decoder and the reset buttons pressed, the decoder is ready to operate. If module one was set up to produce 697 Hz and 1209 Hz, the decoder should activate the station number 1 LED on the decoder and the siren or output device should be sounding. To construct a larger system, you could duplicate the sending modules and decoder for a 30-station system by using another transmitter/receiver frequency. The two systems could be tied together as one large system, or one system could be used as a perimeter alarm, while the second system could be used to provide interior or "space" protection.

The remote ID DTMF alarm is relatively inexpensive to construct and it can be adapted to meet many individual alarm applications.

## DTMF transmitter parts list

| Quantity | Part | Description |
|---|---|---|
| 1 | R1 | 1-k$\Omega$, ¼-W, 5% resistor |
| 1 | R2 | 1-M$\Omega$, ¼-W, 5% resistor |
| 1 | R3 | 10-M$\Omega$, ¼-W, 5% resistor |
| 4 | R4, R8, R9, R10 | 100-k$\Omega$, ¼-W, 5% resistor |
| 2 | R5, R13 | 4.7-k$\Omega$, ¼-W, 5% resistor |
| 1 | R6 | 2-k$\Omega$, ¼-W, 5% resistor |
| 1 | R7 | 10-k$\Omega$, ¼-W, 5% resistor |
| 1 | R11 | 50-k$\Omega$ potentiometer (trimmer) |
| 1 | R12 | 6.1-k$\Omega$, ¼-W, 5% resistor |
| 1 | R14 | 24-k$\Omega$, ¼-W, 5% resistor |
| 1 | Rt1 | 10-k$\Omega$ potentiometer (trim) |
| 1 | Rt2 | 10-k$\Omega$ potentiometer (trim) |
| 1 | C1 | 0.01-$\mu$F, 25-$V_{dc}$ capacitor |
| 2 | C2, C9 | 1-$\mu$F electrolytic, 25 $V_{dc}$ |
| 2 | C3, C8 | 0.05-$\mu$F, 25-$V_{dc}$ capacitor |
| 1 | C4 | 4.7-$\mu$F electrolytic, 25 $V_{dc}$ |
| 1 | C5 | 2-$\mu$F electrolytic, 25 $V_{dc}$ |
| 1 | C6 | 200-$\mu$F electrolytic, 25 $V_{dc}$ |
| 4 | C7A, C7, Ct1, Ct2 | 0.1-$\mu$F, 25-$V_{dc}$ capacitor |
| 1 | IC1 | 555 timer |
| 1 | IC2 | 556 dual timer |
| 1 | IC3 | LM324 op amp |
| 1 | IC4 | CD4011 NAND gate |
| 1 | D1 | 1N4001 diode |
| 1 | RLY | 5-V relay, low current (Radio Shack 275-243) |

## DTMF location alarm receiver/decoder parts list

| Quantity | Part | Description |
|---|---|---|
| 4 | R1, R2, R3, R4 | 100-k$\Omega$, ¼-W, 5% resistor |
| 1 | R5 | 60-k$\Omega$, ¼-W, 5% resistor |
| 1 | R6 | 300-k$\Omega$, ¼-W, 5% resistor |

| | | |
|---|---|---|
| 1 | R7 | 2-k$\Omega$, ¼-W, 5% resistor |
| 1 | R8 | 24-k$\Omega$, ¼-W, 5% resistor |
| 1 | Rt1 | 1-M$\Omega$, ¼-W, 5% resistor |
| 15 | RA thru RO | 2.2-k$\Omega$, ¼-W, 5% resistor |
| 15 | RA1 thru RO1 | 1-k$\Omega$, ¼-W, 5% resistor |
| 2 | C1, C2 | 10-nF, nonpolar, 200-$V_{dc}$ capacitor |
| 1 | C3 | 50-$\mu$F, 25-$V_{dc}$ electrolytic |
| 4 | C4, C5, C6, C7 | 0.1$\mu$F, 25-$V_{dc}$ capacitor |
| 1 | C8 | 1-$\mu$F, 25-$V_{dc}$ electrolytic |
| 1 | C9 | 0.05-$\mu$F, 25-$V_{dc}$ capacitor |
| 1 | C10 | 4.7-$\mu$F, 25-$V_{dc}$ electrolytic |
| 1 | Ct1 | 10-$\mu$F, 25-$V_{dc}$ electrolytic |
| 1 | D1 | 1N4001 diode |
| 1 | Q1 | 2N3904 transistor |
| 15 | SCR1–SCR15 | NTE5405 SCR |
| 1 | X1 | 3.579-MHz color crystal |
| 15 | LED1–LED15 | Red LED |
| 1 | S | Sonalert or piezo siren |
| 1 | RLY1 | 5-V low-current relay (Radio Shack 275-243) |
| 1 | S1 | SPST ON-OFF switch |
| 1 | S2 | NC push-button reset switch |
| 1 | IC1 | 8870 tone decoder |
| 1 | IC2 | CD4514 IC |
| 1 | IC3 | 7400 IC |
| 1 | IC4 | 555 timer |
| 1 | IC5 | LM340-5 IC |

## Portable alarm

A portable alarm can protect your home or cottage while you are away, and it reports directly to a friend or neighbor up to a few miles away. Now you can feel secure in knowing that your property is being monitored, whether you are away for a few hours or many months. The portable alarm is a wireless alarm system that monitors your home with an advanced passive infrared sensor with coverage up to 50 feet. The sensor activates a logic circuit that turns on an FM transmitter, sending a warble tone for 15 seconds to alert your neighbor. A sensitive electret microphone is switched into the circuit to allow your friend to "listen in" for about 4 minutes, at which time he/she can call the police or investigate personally. The portable alarm can be used with an FM transmitter, either an FM broadcast type or a VHF public service band type operating on the itinerate frequencies of 154.570, 151.625, or 154.60 MHz. Your neighbor would use a suitable FM receiver or scanner to monitor your portable alarm.

The circuit in Fig. 8-16 begins with an Eltec 5192 dual-element sensor. The pyroelectric sensor is constructed from a thin film of lithium tantalate, with conductive electrodes attached to

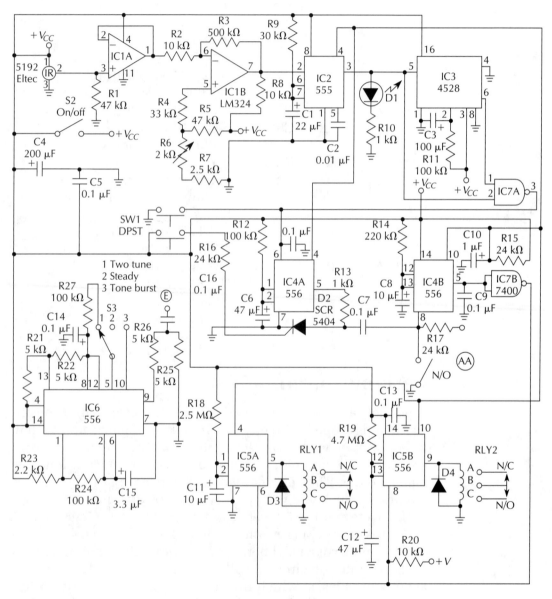

**Fig. 8-16** Portable alarm.

either end of the sensing elements. The pyroelectric material produces an internal electrical field that is collected via the electrodes. Monitoring the charge across the lithium tantalate requires a high-impedance amplifier. Therefore, most sensors are packaged with an FET amplifier, a load resistor, and an impedance converter. These detectors are very sensitive and respond quickly to infrared intensity changes. Both elements are mounted in a TO-5 can with a square aperture fitted with an infrared filter. The dual-element detector is wired parallel-opposed, which helps to eliminate the chances of false alarms in the system.

A person walking in the field of view causes two distinctive outputs, one negative and one positive. The detector produces no output when both elements receive equal radiation, but as soon as one element receives slightly more radiation than the other, an output is produced. In practical applications, it is necessary to focus the IR radiation onto the sensing elements. A Fresnel lens is generally placed in front of the sensors, as shown in Fig. 8-17. The low-cost Fresnel lens comes in many different configurations. One of the two most popular lenses is a long-range type, model LR-V1, which will "see" out to 50 feet, but with a narrow beamwidth. The other lens, a model WA-V3, is a wide-angle lens that "sees" out to 12 feet with a wide-angle coverage. Figure 8-18 illustrates both the wide-angle beam coverage and the narrow beamwidth coverage.

The Eltec 5192 sensor is a three-lead device. Pin 1 is the power-supply input, pin 2 is the signal output, and pin 3 is the case or ground pin. The circuit begins with R1, a 47-k$\Omega$ resistor that draws current for the sensor element. The signal output from the pyroelectric sensor on pin 2 is fed to an LM324, a single-sup-

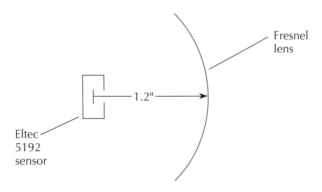

***Fig. 8-17*** *Portable alarm sensor-to-lens diagram.*

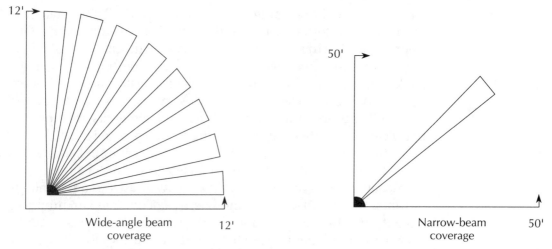

*Fig. 8-18* Portable alarm narrow/wide beam coverage.

ply quad op amp. IC1A isolates the sensors, using a simple voltage follower. The signal is then passed onto IC1B, which acts as a signal amplifier. R3 sets the overall gain of the amplifier. The network connected to pin 5 of IC1B is used as a threshold control and is adjusted via R6. The output from pin 7 of IC1B passes to a 555 monostable multivibrator, which produces a time pulse of about 1 second. The LED D1 assists in the setup and adjustment of R6. The signal from pin 3 of IC2 feeds IC3, a CD4528, which acts as a discriminator circuit. The time period of the window is set by the C3/R11 combination for about 15 seconds, to pass a few pulses from IC2 through the CD4528 before resetting. This window detector helps to prevent false alarms. The 7400 NAND gate inputs the signals from IC2 and IC3 to complete the discriminator.

The signal passes to the heart of the alarm system, IC4, a 556 dual timer. The first half of IC4A is the exit/entry timer, which allows you to leave the room when setting the alarm. This delay is also used when reentering the room to turn off the alarm unit. The exit/entry timer gives you about 15 seconds, which is ample time to leave the protected space. The C6/R12 combination sets this period. Once the exit/entry timer has been started via S1, the SCR D2 is fired, thus preventing IC4B from being triggered during the delay period.

Multiple sensors can be wired to point AA, pin 8 of IC4B. Any normally open sensor, such as a floor-mat or vibration sensor, can be connected to this point.

The time period of IC4B is set by the C8/R14 pair. The output of IC4B is a positive-going pulse and it must be inverted to trigger IC5. IC5A and IC5B are two independent timers. The first timer, IC5A, switches between the warble-tone generator and the electret microphone used to "listen in" to the protected area. The C11/R18 combination is set to about 15 seconds. Once the alarm is triggered, the warble tone is transmitted for 15 seconds to get the attention of your neighbor.

At the same instant, IC5A is triggered. IC5B also begins its timing period and power is applied to the FM transmitter module. See Fig. 8-19 for relay connections to both the warble-tone generator and the transmitter. The timer at IC5B is set up to allow the transmitter to send a signal for about 4 minutes, after which time the transmitter is turned off, providing there is no further triggering of the pyroelectric sensor. A power-on reset circuit is formed by C10/R15, which resets all timer on power-up.

The warble-tone generator is comprised of IC6, a dual-timer IC set up in the astable mode. Switch S3 selects the three possible tone outputs. The C14/R22 and the R24/C15 combinations determine the actual frequencies generated by the warble-tone generator. The output of the sound generator at point E is coupled via C16, a 0.1-µF capacitor connected to the audio input of the transmitter module.

The next section discusses the transmitter portion of the portable alarm. I recommend that you purchase an FM transmit-

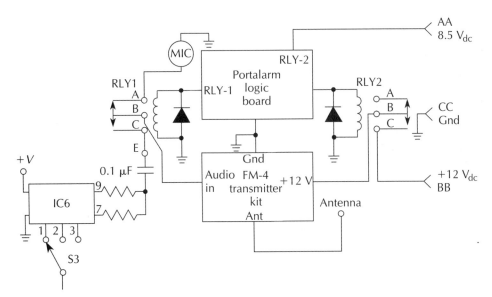

*Fig. 8-19* *Portable alarm board interconnection diagram.*

ter or exciter module, because the building and testing of rf transmitters is not a trivial pursuit, and usually requires special equipment for testing and adjustments. A few alternatives are suggested for obtaining a transmitter for this project. One approach would be to use a single CB or VHF walkie-talkie, or perhaps a surplus cordless-telephone headset, which transmits on 46 MHz. Any of the above transmitters could be used with a scanner. Another approach would be to purchase a small wireless FM transmitter (listed in the appendix) and perhaps change the frequency as desired. An FM-4 miniature FM transmitter kit is available from Ramsey Electronics (see Fig. 8-20). This low-cost broadcast-band FM transmitter has a range of ½ mile. The frequency can be easily modified if you do not wish to use it on the FM broadcast band. A scanner has a squelch control to eliminate noise when there is no signal.

Two schematics are presented for you to consider. The first transmitter is a low-power FM-4 kit schematic. When used with a dipole antenna, it has a ½ mile range. A yagi antenna increases the range even further. The second transmitter schematic is for the FM-5 (see Fig. 8-21), a 5-W FM broadcast unit. It too can be modified for higher frequency use. We do not recommend using the FM-5 if you live in a suburban area, because you would not need 5 W of power. The FM-5 is recommended for rural use or

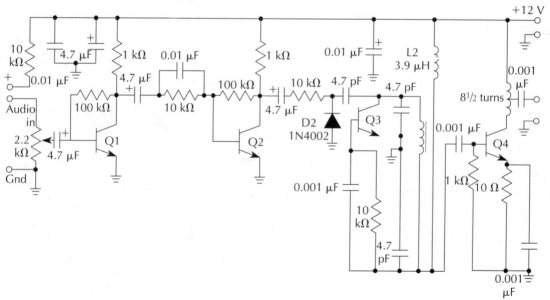

**Fig. 8-20** *Portable alarm FM-4 transmitter option 1.*

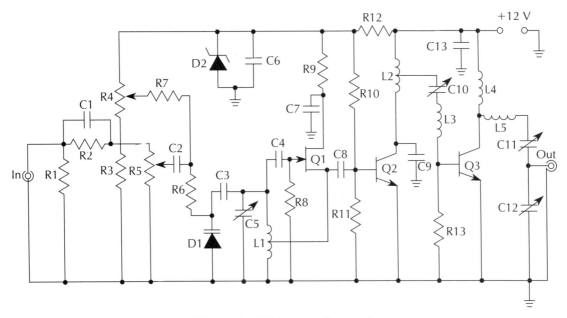

**Fig. 8-21** *FM-5 transmitter option.*

where the portable alarm is placed a long distance from a neighbor's house, such as in the country or in the woods. If the transmitter is used above the FM broadcast band, the frequencies of 151.625, 154.57, and 154.66 are suggested, because they are reserved for itinerate communications.

The power for the portable alarm is provided by a battery charger (see Fig. 8-22), which charges a 12-V 1.2-A/hr gell-cell battery. A regulator powers the logic circuit and sensor. The transmitter is powered directly from the 12-V battery. In this configuration, the portable alarm remains completely wireless. Someone tampering with the power line is no longer a threat, because the battery is always ready to provide power and is continuously charged.

Construction of the portable alarm is somewhat involved. The logic board is constructed on a G-10 glass-epoxy circuit board measuring 5×6 inches. The logic board, transmitter, and power supply are all placed inside of a chassis box measuring 8×10 inches. The transmitter module should be shielded from the logic board and power supply. The transmitter board should be placed inside of a box of its own, contained in the system box. The most difficult part of construction involves mounting the Eltec sensor. Refer to Fig. 8-23 for mounting schemes. A Fresnel lens must be

254  Unique High-tech Security Projects

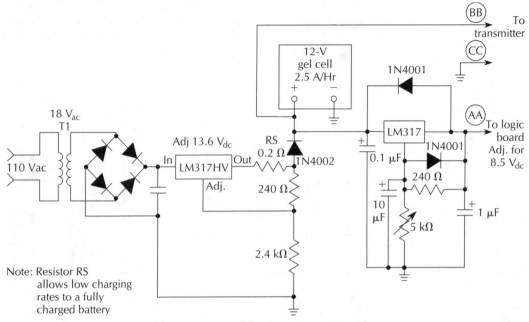

**Fig. 8-22**  *Portable alarm power supply.*

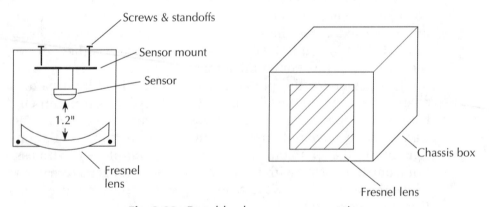

**Fig. 8-23**  *Portable alarm sensor mounting.*

mounted 1.2 inches in front of the pyroelectric sensor, as shown in Fig. 8-17, and bent into an arc. Enclose the sensor in a small metal box so that air currents and rf energy will not affect the sensor. One approach is to mount the sensor in a tuna can, with the plastic Fresnel lens placed over a cutout on the side of the can. Mount the detector near the center of the can, 1.2 inches from the lens. The tuna can could then be mounted on top of the system box. A sensor coverage diagram in Fig. 8-18 depicts both the long- and short-range coverage models.

The front panel of the system chassis contains S2, the ON-OFF switch, the exit/entry push button S1, and the tone-selector switch S3. Mount LED D1 and potentiometer R6 on the front panel. Mount the antenna jack, external sensor jacks, and power cord at the rear of the chassis box.

Operation of the portable alarm is straightforward. First, select the warble tone that will be transmitted via S3. Attach your antenna to the transmitter. You can construct a half-wave dipole antenna (length in feet equals 492 divided by the frequency in megahertz), or you can use a whip antenna, but with reduced range. You may provide additional open-circuit sensors or switches, such as magnetic door switches, floor mats, or vibration sensors, to point AA on the logic board. The portable alarm should be placed on a counter or table with the longest possible nonobstructed view of the sensor to "see" across your home or cabin. Never point the sensor into a fire or at direct sunlight.

Turn on a frequency-compatible receiver to be used with your portable alarm. Now, turn on the portable alarm unit, and adjust R6 to a point just below lighting LED D1. Wave your hand in front of the sensor, and the LED should light. The transmitter should begin sending the warble tones for about 15 seconds. After the 15-second time period, the electret microphone should switch to the transmitter. To test the exit/entry function, turn off the portable alarm for a few seconds, then turn S2 on and press S1. Quickly walk away from the system to ensure the unit does not trigger. Wait about 30 to 40 seconds and enter the protected area. The alarm should trigger. The exit/entry timer will also allow you 15 seconds to return into the protected area to turn the unit off, preventing the alarm from sounding.

The portable alarm now is ready to protect your home or cabin. You can feel secure in knowing that the portable alarm is standing sentinel, 24 hours a day.

## Portable alarm parts list

| Quantity | Part | Description |
|---|---|---|
| 2 | R1, R5 | 47-k$\Omega$, ¼-W, 5% resistor |
| 3 | R2, R8, R20 | 10-k$\Omega$, ¼-W, 5% resistor |
| 1 | R3 | 500-k$\Omega$, ¼-W, 5% resistor |
| 1 | R4 | 33-k$\Omega$, ¼-W, 5% resistor |
| 1 | R6 | 2-k$\Omega$ trim potentiometer |
| 1 | R7 | 2.5-k$\Omega$, ¼-W, 5% resistor |

| Quantity | Part | Description |
|---|---|---|
| 1 | R9 | 30-k$\Omega$, ¼-W, 5% resistor |
| 2 | R10, R13 | 1-k$\Omega$, ¼-W, 5% resistor |
| 3 | R11, R12, R24 | 100-k$\Omega$, ¼-W, 5% resistor |
| 1 | R14 | 220-k$\Omega$, ¼-W, 5% resistor |
| 3 | R15, R16, R17 | 24-k$\Omega$, ¼-W, 5% resistor |
| 1 | C1 | 22-$\mu$F electrolytic, 25 $V_{dc}$ |
| 1 | C2 | 0.05-$\mu$F 25-$V_{dc}$ capacitor |
| 1 | C3 | 100-$\mu$F electrolytic, 25 $V_{dc}$ |
| 1 | C4 | 200-$\mu$F electrolytic, 25 $V_{dc}$ |
| 5 | C5, C7, C13, C14, C16 | 0.1$\mu$F, 25-$V_{dc}$ |
| 2 | C6, C12 | 47-$\mu$F electrolytic, 25 $V_{dc}$ |
| 2 | C8, C11 | 10-$\mu$F electrolytic, 25 $V_{dc}$ |
| 1 | C10 | 1-$\mu$F electrolytic, 25 $V_{dc}$ |
| 1 | C15 | 3.3-$\mu$F electrolytic, 25 $V_{dc}$ |
| 1 | IR | Eltec 5192 dual-opposed pyroelectric sensor |
| 1 | D1 | LED |
| 1 | D2 | SCR5404 SCR |
| 2 | D3, D4 | 1N4002 diode |
| 1 | R18 | 2.5-M$\Omega$, ¼-W, 5% resistor |
| 1 | R19 | 4.7-M$\Omega$, ¼-W, 5% resistor |
| 4 | R21, R22, R25, R26 | 5-k$\Omega$, ¼-W, 5% resistor |
| 1 | R23 | 2.2-k$\Omega$, ¼-W, 5% resistor |
| 1 | IC1 | LM324 op amp |
| 1 | IC2 | 555 timer |
| 1 | IC3 | CD4528 IC |
| 3 | IC4, IC5, IC6 | 556 dual timer |
| 1 | IC7 | 7400 NAND gate |
| 1 | RLY1, RLY2 | 5-V low-current relay (Radio Shack 275-243) |
| 1 | SW1 | DPST push-button switch |
| 1 | SW2 | SPST ON-OFF switch |

## FM-4 kit parts list

| Quantity | Part | Description |
|---|---|---|
| 2 | Q1, Q2 | 2N3904 transistor |
| 2 | Q3, Q4 | 7001 IC |
| 1 | L1 | 8½ turns, 24-gauge enameled close-spaced |

⅜-inch diagonal form, tap at 2 turns from 12-V side

| | | |
|---|---|---|
| 1 | L2 | 3.9-µH molded choke |
| 1 | L3 | 5-turn slug-tuned coil |
| 1 | D1 | 1N4002 diode |

NOTE: All resistors are ¼-W, 5% carbon. Values are shown on the schematic. All capacitors are listed on schematic.

## 5-W FM transmitter parts list

| Quantity | Part | Description |
|---|---|---|
| 1 | R1 | 10-kΩ, ½-W, 5% resistor |
| 1 | R2 | 15-kΩ, ¼-W, 5% resistor (for music input; short or 0Ω for voice input) |
| 1 | R3 | 12-kΩ, ½-W, 5% resistor |
| 1 | R4 | 100-kΩ potentiometer (trim) |
| 1 | R5 | 500-Ω potentiometer |
| 3 | R6, R7, R8 | 68-kΩ, ½-W, 5% resistor |
| 1 | R9 | 150-Ω, ½-W, 5% resistor |
| 1 | R10 | 22-kΩ, ½-W, 5% resistor |
| 1 | R11 | 1-kΩ, ½-W, 5% resistor |
| 1 | R12 | 100-Ω, 1-W, 5% resistor |
| 1 | R13 | 56-Ω, ½-W, 5% resistor |
| 1 | C1 | 0.005-µF, 25-$V_{dc}$ capacitor |
| 1 | C2 | 10-µF, 25-$V_{dc}$ electrolytic |
| 3 | C3, C4, C8 | 12-pF, NPO, 25-$V_{dc}$ capacitor |
| 1 | C5 | 3–30-pF, 25-$V_{dc}$ trimmer capacitor |
| 1 | C6 | 47-µF, 25-$V_{dc}$ electrolytic |
| 2 | C7, C13 | 0.1µF, 25-$V_{dc}$ disk capacitor |
| 2 | C9, C10 | 9–50-pF, 25-$V_{dc}$ trimmer capacitor |
| 2 | C11, C12 | 16–100-pF, 25-$V_{dc}$ trimmer capacitor |
| 1 | D1 | MV2113 capacitive diode |
| 1 | D2 | 1N5240 10-V zener diode |
| 1 | Q1 | 2N4416 FET |
| 1 | Q2 | 2N3866 or 2N4427 npn transistor |
| 1 | Q3 | MRF237 power transistor |
| 1 | L1 | 6 turns, tapped at 2½ turns |
| 1 | L2 | 5 turns, center-tapped |
| 1 | L3 | 4-turn coil |
| 1 | L4 | 2½-turn coil |
| 1 | L5 | 4½-turn coil |

All coils solid 22-gauge on ¼-inch form
Heat sinks required for Q2, Q3

## Storm warn

Electrical storms are responsible for more damage and destruction to sensitive electronic equipment than any other source known. During an electrical storm, tremendous electrical surges race across the sky and plummet to the ground in a fraction of a second, often destroying sensitive equipment in their path. Electrical equipment is not always destroyed by a "direct hit," but often due to poor grounding. Secondary or residual paths "trap" sensitive electronic equipment, causing particular parts of a circuit to be destroyed.

The storm warn is a unique project that can protect computers, radio equipment, and other types of sensitive electronic instruments from the ravages of electrical storms (see Fig. 8-24). The storm warn automatically disconnects your electronic equipment from the power line. The storm warn could also disconnect antenna systems from radio gear during an electrical storm.

The storm warn is designed to give an advance audible warning of an approaching storm up to 15 miles away. This advance warning allows you to close files or "spin down" a large hard drive, before the system shuts down your computer/radio gear for 15 to 20 minutes or longer, which it does if it is continuously triggered. The storm warn detects electrical storm activity in the 500-kHz radio spectrum. Electrical storms generate enormous amounts of wideband rf noise and transients in the low-frequency spectrum below the AM broadcast band. The storm warn accumulates and integrates these storm transients. The pulses are passed to a time-window discriminator, so that only a certain number of pulses are allowed to pass in a finite time period for the storm warn to trigger.

The storm warn allows you 2 to 4 minutes, once triggered, before the system automatically disconnects your computer from the power line/or antenna for 15 to 20 minutes. If additional triggers or events are detected after the initial trigger, the timer will accumulate time and add additional time to the "off time" until no more triggers are detected. The storm warn also incorporates built in "spike" protection during regular operation. Both transverse and common-mode protection are provided by using varistors or Transorbers, V1 through V3 across the power line.

Operation of the storm warn begins with the rf section or receiver portion. The rf detector is essentially an AM radio tuned to 535 kHz. The detector is formed by L1, C1, and D1, a germanium diode. The detected signal is coupled to IC1, a 741 op amp.

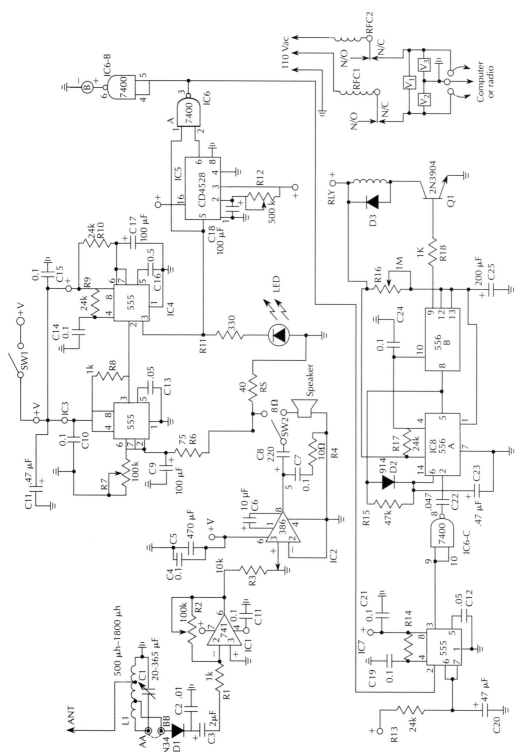

**Fig. 8-24** Storm warn.

Resistor R2 sets the op-amp gain. The trimmer can be replaced with a 100-Ω resistor. The output of IC1 is fed to IC2 via R3, a 10-kΩ potentiometer. The output of IC1 is further amplified by IC2, an LM386 amplifier with a gain of 200. The output of IC2 is connected to S2, allowing you to switch to an 8-Ω speaker, which greatly aids in testing the storm warn system.

Under normal operation, S2 is switched to R6. Coil L1 can be obtained from an old transistor radio. The coil used in the prototype was a four-lead (two-tap) version. The coil should have a value of 500–1800 µH with an adjustable ferrite tuned core.

An antenna was constructed from an 8 to 10-foot length of RG-174 coax. The antenna is connected to L1, as shown in Fig. 8-25. The last 4 feet of coax nearest the free end has the shield portion removed, and it acts as the active part of the antenna (more on this topic later).

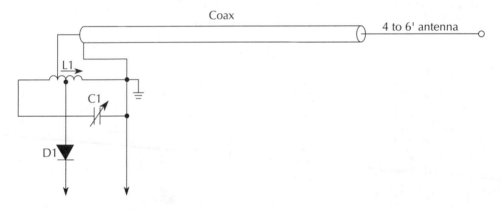

**Fig. 8-25** *Storm-warn antenna hook up.*

Resistor R6 couples the receiver portion of the system to IC3, the charge accumulator/trigger. Static pulses are integrated over time via C9. Resistor R7 controls the sensitivity of the trigger section. The output of IC3 is coupled to IC4, a one-shot that shapes the output pulses. The pulse duration is controlled by R10/C17. A power-on reset circuit is formed by R9/C14, to prevent triggering when powering the system. An LED indicator is connected to the output of IC3 to display the trigger events as they occur.

The output of IC4 is fed to IC5, the "window discriminator," a CD4528 CMOS one-shot. The pulses from IC4 and IC5 are NANDed together at IC6, forming the time window or pulse discriminator. No output at IC6 will occur unless more than one trigger or pulse has arrived from IC4 during the active time of

IC5. The time period of IC5 can be adjusted by R12/C18. This adjustment may require some tinkering.

When a valid event is determined, IC6 will sound an electronic buzzer for 2 to 4 minutes, adjustable. This 2-minute timer allows you enough time to spin-down a disk or close your files if you are near your computer. This time period can be altered by R13/C20 at IC7. The 556 dual timer at IC8 is configured as a retriggerable timer. If the storm-warn system "sees" any additional triggers once the system has shut down the output device or computer, then the "off time" is increased for an additional time period or until no further triggers are detected.

The "off timer" can be set from 10 to 20 minutes by changing the values R16/C25. Once no further triggers have been detected and IC8 times out, power is restored to the computer or equipment.

Under normal operation, power varistors V1 through V3 or Transorbers protect your equipment from common-mode and transverse spikes, as well as transients. Your equipment is always being protected, thus helping to ensure a long life to your computer/radio gear.

The storm-warn circuit can be powered in three different ways, as shown in Figs. 8-26, 8-27 and 8-28. The 741 op amp requires a plus-and-minus supply, but the minus supply requires

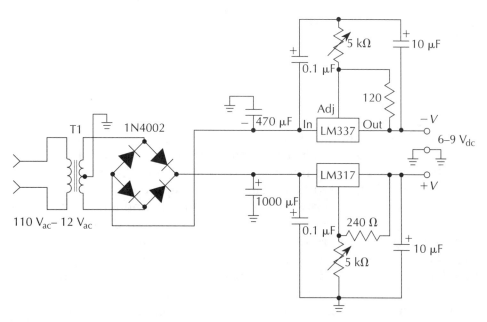

*Fig. 8-26  Storm-warn power supply.*

262   Unique High-tech Security Projects

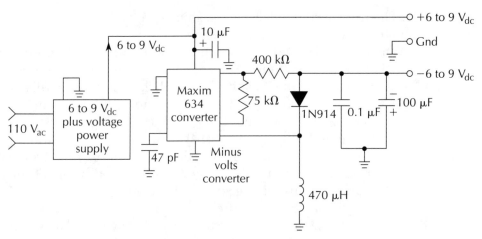

**Fig. 8-27** *Storm-warn minus voltage converter.*

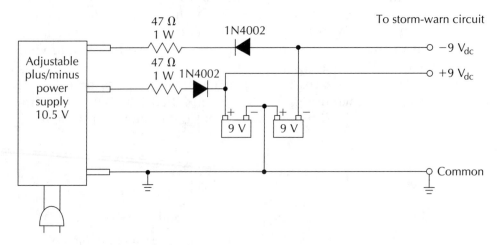

**Fig. 8-28** *Storm-warn rechargeable power supply.*

very little current. Figure 8-26 illustrates a simple plus-and-minus supply that can be used with the storm-warn circuit. The dual supply uses the familiar LM317/LM337 adjustable regulators. This supply is easy to fabricate. Another approach is to use a single plus-voltage supply with a minus-voltage generator such as the Maxim 634, as shown in Fig. 8-27. This chip produces the required minus low-current supply voltage necessary to power the 741. The power source shown in Fig. 8-28 is the recommended power supply, using the dual-voltage supply shown in Fig. 8-26. The supply charges two small 6-V, 1.2-A, gel-cell batteries. In this scheme, the batteries are always being trickle-charged. In the event of a storm, the batteries will keep the storm warn in operation, ensuring that your computer/radio equipment remains totally disconnected from power for the duration of the damaging storm. If you can locate a plus/minus surplus wall-cube supply, you could use that to charge the gel-cell batteries.

Construction of the storm warn is relatively straightforward. A 6 × 6-inch G-10 glass-epoxy 1-oz copper circuit board was used for the prototype unit. An aluminum box measuring 10×5 inches was used to house the storm-warn circuit, power supply, and batteries. Two switches, S1 and S2, were both mounted on the front of the chassis box. The speaker, three-pronged outlet, power cord, and antenna jack were mounted at the rear of the chassis. Pay careful attention to the polarity of the capacitors and the diodes, especially in the power-supply circuits. Observe diode D1 polarity as well as the polarity of Q1, the relay driver. Coil L1 can be mounted in a plastic cable clamp or any nonmetallic holder.

Calibration and testing of the storm-warn system requires some extra care and attention to ensure reliable operation during a storm. Attach a ground wire and an antenna to L1. Next, switch S2 to the speaker position and apply power via S1. The receiver is essentially an AM crystal radio with high amplification. The highly amplified rf section detects rf-transient noise during an electrical storm. Tune C1 and L1 to a station to ensure that the rf section is working properly. Now, tune the receiver to the lower portion of the band around 535 kHz. Make sure it is not tuned to a station in your area, because you want to detect only background noise during regular operation.

Determine the best placement of the storm warn with respect to the computer. Since the computer usually generates noise itself, orient the storm warn and antenna away from your computer. Listen on the storm-warn speaker for the most quiet positioning, to ensure that the computer will not interfere with the storm-warn circuit. The coax antenna described earlier should be placed near an outside window, if possible. Now apply +6 V to R6 at the terminals of switch S2. You should see the LED at IC4 light up.

Next, proceed to test IC5 and IC6. Listen for the buzzer to sound. Once again, trigger the circuit by applying 6 V to R6. Wait about 10 seconds the CD4528 time-out period, and apply another trigger to R6. The buzzer should now sound. Try this a few times to ensure that this portion of the circuit is working properly.

IC7 has a delay time of 2 to 4 minutes, for you to close your computer files if you are present, before your computer is shut down. Once the buzzer has sounded, look for an output on pin 3 of IC7 with an oscilloscope or LED.

If everything is working, proceed to the last stage of the circuit, IC8, the retriggerable system timer that shuts the computer system down for the 10 to 15-minute time period during storm activity. Again trigger R6, to sound the buzzer. Wait 2 to 4 minutes, and then Q1 should activate relay RLY. The relay RLY could control a second power relay. The system "off timer" at IC8 can be adjusted from 10 to 20 minutes, as described earlier.

Your storm-warn system should now be operating as intended. Make sure that your computer will not interfere with the storm-warn receiver section, as discussed earlier. At last, the system is ready to protect your electronic equipment. Switch S2 to disable the speaker. The rf section will now be connected to the timer/controller section. A final test for the storm warn is to bring a soldering gun near the circuit and rapidly turn the gun off and on a few times. The storm warn should now come to life. Once you are satisfied that the storm warn is working correctly, you can plug your computer into the outlet you have mounted on the storm-warn chassis. You can feel a little more secure in knowing that the storm warn is an ever-present guardian, watching over your electronic equipment.

## Storm-warn parts list

| Quantity | Part | Description |
| --- | --- | --- |
| 3 | R1, R8, R18 | 1-kΩ, ¼-W resistor |
| 1 | R2 | 100-kΩ, ¼-W resistor or trim pot |
| 1 | R3 | 10-kΩ trim pot |
| 1 | R4 | 10-Ω, ¼-W resistor |
| 1 | R5 | 40-Ω, ¼-W resistor |
| 1 | R6 | 75-Ω, ¼-W resistor |
| 1 | R7 | 100-kΩ trim pot |
| 3 | R9, R10, R13 | 24-kΩ, ¼-W resistor |
| 2 | R14, R17 | 24-kΩ, ¼-W resistor |
| 1 | R11 | 330-Ω, ¼-W resistor |
| 1 | R12 | 220-kΩ, ¼-W resistor |
| 1 | R15 | 47-kΩ, ¼-W resistor |
| 1 | R16 | 1-MΩ trim pot |
| 1 | C1 | 20–365-pF variable capacitor |
| 1 | C2 | 0.01-μF, 25-$V_{dc}$ capacitor |
| 1 | C3 | 2.2-μF, 25-$V_{dc}$ capacitor |
| 4 | C4, C7, | 0.1-μF, 25-$V_{dc}$ capacitor |

| Quantity | Part | Description |
|---|---|---|
| 4 | C10, C14, C15, C19, C21, C24 | 0.1-μF, 25-$V_{dc}$ capacitor |
| 1 | C5 | 470-μF, 25-$V_{dc}$ electrolytic capacitor |
| 1 | C6 | 10-μF, 25-$V_{dc}$ electrolytic capacitor |
| 2 | C8, C25 | 220-μF, 25-$V_{dc}$ electrolytic capacitor |
| 3 | C9, C17, C18 | 100-μF, 25-$V_{dc}$ electrolytic capacitor |
| 3 | C12, C13, C16 | 0.05-μF, 25-$V_{dc}$ capacitor |
| 2 | C11, C20 | 47-μF, 25-$V_{dc}$ electrolytic capacitor |
| 1 | C22 | 0.047-μF, 25-$V_{dc}$ capacitor |
| 1 | C23 | 4.7-μF, 25-$V_{dc}$ electrolytic capacitor |
| 1 | D1 | 1N34 germanium detector diode |
| 2 | D2, D3 | 1N914 silicon diodes |
| 1 | LED | Red LED |
| 3 | V1, V2, V3 | 120-$V_{ac}$ varistors or transorbers |
| 1 | IC1 | 741C Op amp |
| 1 | IC2 | LM386 audio amplifier |
| 3 | IC3, IC4, IC7 | 555 general-purpose timers |
| 1 | IC5 | CD4528 monostable multivibrator |
| 1 | IC6 | 7400 NAND gate |
| 1 | IC8 | 556 dual timer |
| 1 | RLY | 6–9-V DPDT relay (5–10-A contacts) |
| 1 | Q1 | 2N3904 npn transistor |
| 1 | SW1 | SPST on-off switch |
| 1 | SW2 | SPDT speaker switch |
| 1 | L1 | 500–1800-μH coil with ferrite adjustable core |

# Video sentry

The video sentry is a unique, new, security/surveillance system, which allows you to "view" a distant room or location, as well as permits you to "listen" to that same location using a sensitive audio monitoring circuit. In addition, you can remotely control a number of appliances, recorders, or sounding/warning devices via the public telephone network.

The video sentry consists of two independent station setups. The remote sensing or pickup station consists of a videophone kit, a black-and-white camera, and the video sentry decoder/control circuit. The monitoring or display station consists of a videophone kit, a monitor, and a Touch-Tone phone, or a rotary phone with an external DTMF tone generator. The video sentry system diagram is shown in Fig. 8-29.

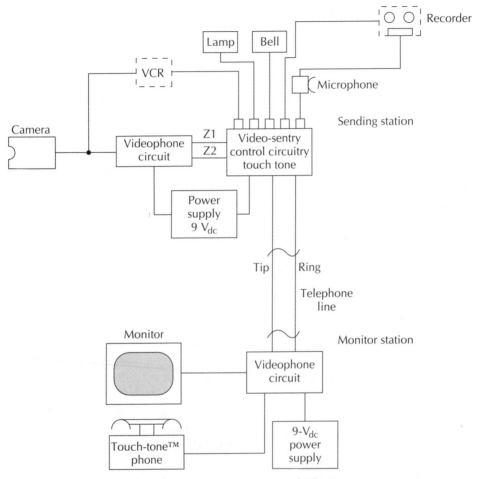

**Fig. 8-29** *Video sentry system diagram.*

Simply pick up a Touch-Tone phone and dial the number of the phone line where the video sentry was installed, such as your home, office, or warehouse. As you finish dialing the last digit of that phone number, press the star (*) button on the tone pad. The remote video sentry station will automatically answer the remote phone line before the phone even rings. The videophone circuit begins to send still video pictures of the remote location for up to 5 minutes, or longer if desired. The video sentry also allows unlimited audio listening of the remote site, as well as remote control of recorders, warning devices, or lamps.

The video sentry is a modern-day infinity transmitter that

can be used anywhere across any geographic distance. The video sentry can be used to monitor your office while you are away, or to watch your children or babysitter. It could even be used to watch your neighborhood. The video sentry could also be used to keep tabs on old folks living alone. Your imagination might also provide additional applications for the video sentry.

The heart of the videophone is an amazing wonder of integration and sophistication, providing high-resolution still pictures with greater than 50,000 pixels and up to 50 gray-scale levels. This videophone circuit does indeed provide the best video of any videophone available to date. The large number of gray-scale levels provides the increased appearance of fine detail and preserves the original video picture quite nicely. The videophone sends a 200×242½-line picture with a proprietary pulse-width-modulation scheme in the 1700 to 3440-H audio range.

The videophone circuit consists of the videophone chip, a flash A/D converter, a memory chip, an output buffer, a speed circuit, and a handful of discrete components. The videophone is available in a 4-bit or 6-bit conversion configuration. The circuit shown in Fig. 8-30 illustrates the 4-bit A/D version. Building the 6-bit version for increased picture quality.

The manufacturer has recently announced a color version of the videophone, using a standard NTSC camera or VCR input. PMC, the manufacturer of the videophone, also has announced a printing adaptor as well as an increased memory version. The basic 6-bit unit with black-and-white conversion is quite adequate for this application. The basic videophone can be purchased as a kit, or as a wired, completely ready-to-operate unit, or you can purchase individual parts from PMC, listed in the Appendix.

The videophone is capable of four modes of operation—send, receive, capture, and autosend. For our application, the autosend function is hardwired, so no external switches are needed to operate the videophone. The autosend feature allows the videophone to capture and automatically send a still picture every 30 seconds, for an indefinite time period. The picture is sent every 8 to 12 seconds. The additional time is used to capture the picture from a fast-scan camera and to allow the necessary handshake requirements between transceivers.

The main videophone circuit is shown in Fig. 8-30, and illustrates the videophone chip, the flash A/D converter, clock, output buffer, and phone coupler. Figure 8-31 shows the video input and output circuitry that supports the black-and-white

268  Unique High-tech Security Projects

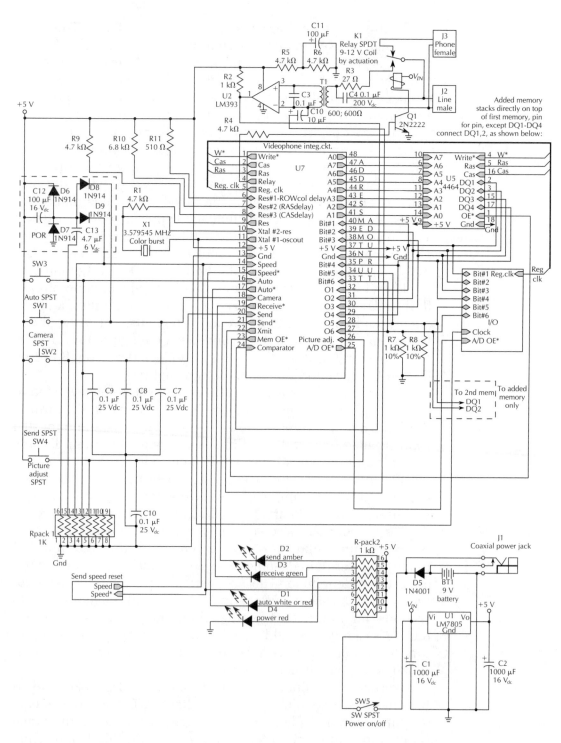

**Fig. 8-30** *Videophone #1 main circuit.*

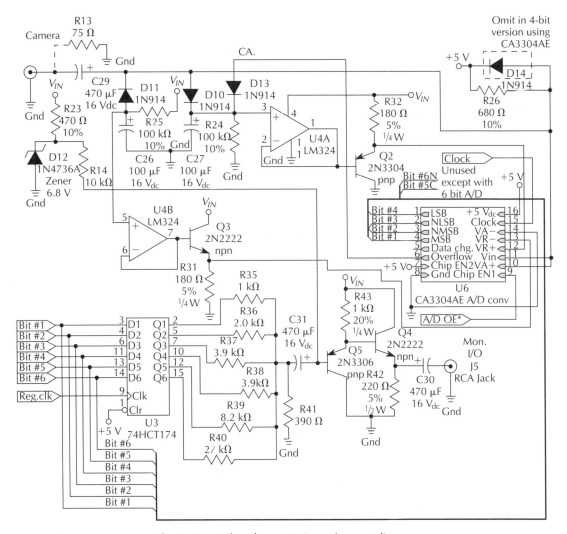

**Fig. 8-31** *Videophone #2 input/output diagram.*

camera, a 1-V$_{p-p}$ standard NTSC unit. The output section drives a composite black-and-white monitor, a modified computer TTL monitor, or, with the addition of an rf modulator, a television set. Figure 8-32 depicts the speed circuit that controls the transmission speed of the system.

The decoder/control circuitry of the video-sentry system is shown in Fig. 8-33. The circuit begins with a Teltone 8870 DTMF decoder chip. A 3.579-MHz color-burst crystal establishes the master oscillator for the tone decoder. The telephone line is coupled to pins 1 and 2 of the decoder chip via two 10-nF nonpolarized capacitors. The resistors connected to pins 3 and 4 establish a voltage

270  Unique High-tech Security Projects

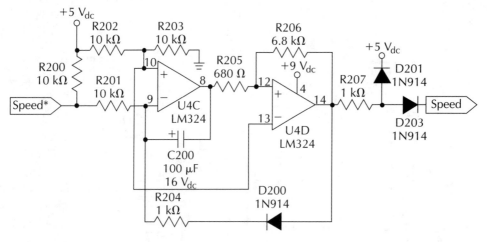

**Fig. 8-32**  *Videophone #3 speed circuit diagram.*

## Video-sentry control system

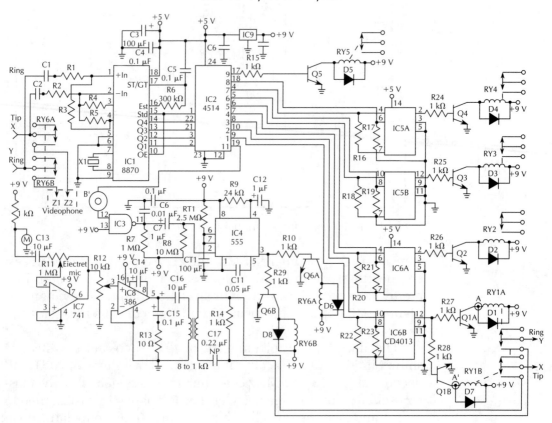

**Fig. 8-33**  *Videosentry system diagram.*

reference for the 8870 input. The STD or pin 15 on the decoder is a delayed logic output, which provides a valid output as a tone pair is received. The standard output is tied to the output enable, pin 10, which is connected to the strobe input, pin 1 of the CD4514. The CD4514 is a 4 to 16-line decoder, which decodes the binary output of the tone decoder and provides 15 separate outputs, representing digits 1 through 10 plus additional control outputs. The star (*) button is decoded and used for the video function. A few other outputs are available but not used, because most Touch-Tone pads will not provide these tones unless modified.

Outputs one through eight are coupled to IC5 and IC6, a pair of CD4013 flip-flops. These flip-flops control the output relays and latch them once a tone is decoded. Pressing digits 1 and 2 latches and unlatches relay one. Relays one through four operate in the same manner. Output nine uses a nonlatching relay, and this relay will only operate until a new digit is decoded. The LM340-5, IC9, is a 5-V regulator, used primarily to power the tone decode, because its maximum rating is 6 V.

The video control function is selected by pressing the star (*) button on the Touch-Tone pad. Pin 19 of the CD4514 is coupled to a 7400 NAND gate, which feeds IC4, a 555 timer. The 555 is configured as a nonretriggerable timer, with the time-out set for 5 minutes by the Rt1/Ct1 combination. A timer triggers relay six, because once the video signal begins, a continuous stream of audio might interfere with any control tone sent during this time. To provide reliable operation, the timer was implemented for the video-mode of operation.

After the 5-minute time-out period, the videophone is disconnected from the phone line. If you wish to activate the audio listening feature, dial the video sentry phone number, and digits 1 and 2 would latch and unlatch this mode. You are permitted to "listen in" for any time period once you select this mode; however, you must manually deselect this mode, once finished, to disconnect the phone line. A highly sensitive electric microphone is coupled to a 741 op amp, which in turn is fed to an LM386 power amplifier. The LM386 is then connected to an 8-$\Omega$ to 1-k$\Omega$ miniature audio matching transformer. A 1-k$\Omega$, 2-W resistor "holds" the telephone line while the audio signal is sent.

The CMOS flip-flops, as mentioned, control most of the functions of the video sentry. Because, the relays connected to the flip-flops are latched, once activated, more than one function can be implemented at a time. You can operate a few devices or appliances simultaneously. Pressing digits 3 and 4 controls the operation of relay two. Digits 5 and 6 activate Q3, which controls relay three. Digits 7 and 8 control relay four. DPDT relays one and six implement the audio and video modes. Miniature SPDT low-current relays RY2 through RY5 must be wired to high-current relays if home appliances are going to be controlled by the video sentry. The low-current relay contacts (normally open) are wired in series to the coil of the high-current relay. The low-current relay switches the 9 to 12-$V_{dc}$ power source to activate the larger relays. The high-current relays switch 110 $V_c$.

The videophone circuitry can be powered by a 9 to 12-V power supply. The videophone and decoder/control board can be powered by a common power supply. Power for the video sentry may be obtained by a few different methods. First, you can construct your own 9 to 12-V supply, or you can use a heavy duty 9-V, 1 to 1½ amp dc wall-cube power supply to provide power to the system. Another approach is simply to purchase a surplus power supply from one of the suppliers in the Appendix.

Additional construction notes include a power jumper placed between pin 14 of IC5 and pin 24 of IC2. Note the three jumpers (J) on the larger decoder board (A). Five connections must be made between the two decoder boards. A power connection for IC power is labeled (+) and a ground connection is marked (−). The wire marked B' goes between the video timer and pin 19 of IC2. Two audio-relay connections labeled A and A' connects IC6 to relays RY1A and RY1B. The last connection between boards is the unregulated relay power supply (+)'.

The videophone and decoder board can be mounted in 6×6×3 inch box. If you obtain a microminiature camera, you could place the camera in the same box, if desired.

Microminiature cameras are readily available for use in model aircraft or railroading. The decoder/control board and the videophone board are mounted one above the other in the chassis box. The exact mounting details will vary, based on the application and type of camera and power supply used. If a power supply is mounted in the same box, keep the camera away from any transformers, because 60-Hz coupling could present problems. No external chassis controls are necessary for the video sentry, because the unit is hardwired for automatic operation, and the control board is always across the phone line, awaiting the control tones. The video sentry is always on, so no power switch is necessary.

The video sentry control board is connected to the videophone board via Z1 and Z2 on relay six. The normally open contacts of RY6 are coupled to the phone jack marked "line" on the videophone card. Power for the videophone is obtained at the input of the regulator IC9.

Operation of the video sentry is quite simple. Simply pick up a Touch-Tone phone and dial the phone number where the video sentry control box is installed. Decide which function you desire, either audio or video. As you finish dialing the last digit of the phone number, select the star (*) button for video or select digit 1 for audio monitoring. The video function, as mentioned, permits video monitoring for up to 5 minutes. The audio mode permits an unlimited listening period.

Start the system in the audio mode. Then you could listen to the area first and determine if the conversation warrants switching to the video mode, which can be selected by pressing the star (*) button and then deselecting the audio mode. Once finished in the audio mode, deselect this function to hang up the telephone line. In the video mode, once activated, the sending station will try to synchronize the receiving videophone station, and this is done automatically.

The videophone at the monitoring station must be set up to allow viewing the distant sending station. Press receive on the videophone. The monitoring-station videophone has two phone jacks on the rear of the circuit board. One is marked "phone." Connect your telephone to this jack. Connect the phone line to the jack marked "line" on the rear of the videophone circuit board. Connect a monitor to the monitor jack and you will now be able to view the distant scenes. A composite video monitor is recommended over a television set, connected via a modulator. Once your monitoring station is set up, you will be able to view an updated picture every 35 seconds, until the time-out period.

The video sentry could easily be used over a dedicated two-wire pair or intercom wire pair. A modified video sentry could operate over a vhf/uhf radio link. This would also make an interesting project. The video sentry is a complex project that should not be attempted by a first-time kit builder. Take time to construct this project in steps. Purchase and build each videophone kit, one at a time. You can call PMC and they will be happy to send or receive a picture to test your videophone. Once the videophone has been built, you can construct the decoder/control board. I hope you enjoy building and operating the video sentry. Perhaps you will find additional applications for this device.

## Video sentry parts list

| Quantity | Part | Description |
| --- | --- | --- |
| 3 | R1, R2, R4 | 100-k$\Omega$, ¼-W, 5% resistors |
| 1 | R3 | 37.5-k$\Omega$, ¼-W, 5% resistor |
| 1 | R5 | 60-k$\Omega$, ¼-W, 5% resistor |

| Quantity | Part | Description |
|---|---|---|
| 1 | R6 | 300-k$\Omega$, ¼-W, 5% resistor |
| 1 | R7 | 1-M$\Omega$, ¼-W, 5% resistor |
| 1 | R8 | 10-M$\Omega$, ¼-W, 5% resistor |
| 1 | R9 | 24-k$\Omega$, ¼-W, 5% resistor |
| 4 | R10, R15, R24 R25, R26, R27, | 1-k$\Omega$, ¼-W, 5% resistor |
| 4 | R28, R29 | |
| 1 | R11 | 1-M$\Omega$ potentiometer |
| 1 | R12 | 10-k$\Omega$ potentiometer |
| 1 | R13 | 10-$\Omega$, ½-W, 5% resistor |
| 1 | R14 | 1-k$\Omega$, 1-W, 5% resistor |
| 1 | Rt1 | 2.5-M$\Omega$, ¼-W, 5% resistor |
| 2 | C1, C2 | 10-nF, 200-V, nonpolarized capacitor |
| 1 | C3 | 100-µF, 25-$V_{dc}$ electrolytic |
| 4 | C4, C5, C10, C15 | 0.1-µF, 25-$V_{dc}$ capacitor |
| 4 | C6, C13, C14, C16 | 10-µF, 25-$V_{dc}$ electrolytic |
| 1 | C7 | 0.22-µF, 25-$V_{dc}$ capacitor |
| 2 | C8, C10 | 0.01-µF, 25-$V_{dc}$ capacitor |
| 1 | C12 | 1-µF, 25-$V_{dc}$ electrolytic |
| 1 | C17 | 0.22-µF, 150-$V_{dc}$ nonpolarized capacitor |
| 1 | Ct1 | 100-µF, 25-$V_{dc}$ electrolytic |
| 8 | RLY1–RLY8 | DPDT low-current relay, 5 V, (Radio Shack 275-243) |
| 4 | RLY2, 3, 4, 5 | SPDT low-current relay, 5 V |
| 6 | Q1–Q6 | 2N3904 |
| 1 | T1 | 8-$\Omega$ to 1-k$\Omega$ transformer |
| 8 | D1–D8 | 1N4001 diode |
| 1 | IC1 | 8870 Teltone decoder |
| 1 | IC2 | CD4514 4 to 16 decoder |
| 1 | IC3 | 7400 NAND gate |
| 1 | IC4 | 555 timer |
| 1 | IC5 | CD4013 flip-flop |
| 1 | IC6 | CD4013 flip-flop |
| 1 | IC7 | LM741 op amp |
| 1 | IC8 | LM386 audio amplifier |
| 1 | IC9 | LM340-5 regulator |

# Appendix

# Suppliers

Ademco Alarms
165 Eileen Way
Syosset, NY 11791
(516) 921-6704
Alarm components/control boxes

Alco Electronics
1830 N. 80th Place
Scottsdale, AZ 85257
rf parts, unusual parts

All Electronics Corporation
P.O. Box 567
Van Nuys CA 91408
(800) 826-5432
New parts, surplus parts

Acqutek Corporation
P.O. Box 187
Sandy, UT 84091
(801) 572-8151
Inexpensive A/D cards

American Design Components
62 Joseph St.
Moonachie, NJ 07074
(201) 939-2710
Surplus parts, power supplies

B Soft Software
444 Colton Rd.
Columbus, OH 43207
A/D cards for PCs

Ball Corp.
3400 Gilchrist Rd.
Moqadore, OH 44260
(216) 784-4456
TV cameras

Circuit Specialists
P.O. Box 3047
Scottsdale, AZ 85275
(602) 966-0764
Kits and parts

Cohu Inc.
5755 Kearny Villa Rd.
San Diego, CA 92123
(619) 277-6700
TV cameras

Consumertronics
2011 Crescent Drive
P.O. Drawer 537
Alamogordo, NM 88310
Parts and overruns

Datak Corporation
3117 Patterson Park Rd.
North Bergen, NJ 07047
(201) 863-7667
PC board supplies

Dicon Alarm Systems
631 Executive Dr.
Willow Brook, IL 60521
Alarm control boxes

Digi-key Corporation
701 Brookes Avenue, South
P.O. Box 677
Thief River Falls, MN 56701
(800) 344-4539
Good selection of parts

Edmund Scientific
Corporation
101 E. Glouseter Pike
Barrington, NJ 08007
(609) 573-6250
Kits, mirrors, laser, telescopes, science supplies

Electronic Energy Control
380 S. Fifth Street
Suite 604
Columbus, OH 43215
(800) 842-7714
Low-cost computer cards

Electronic Goldmine
P.O. Box 5408
Scottsdale, AZ 85261
(602) 451-7454

Figaro Engineering
P.O. Box 355
Wilshire Drive East
Wilmette, IL 60091
(312) 256-3546
Tin-oxide gas sensors

GBC CCTV Corp.
315 Hudson St.
New York, NY 10013
(800) 221-2240
B/W low-light microminiature TV camera

Hamilton Avnet
(800) 962-1302
Very good selection of parts

Hamtronics
65 Moul Road
Hilton, NY 14468
(716) 392-9430
Radio kits

Herbach & Rademan
401 East Eric Ave.
Philadelphia, PA 19134
(800) 848-8001
Surplus items

Information Unlimited
P.O. Box 719
Amherst, NH 03031
(800) 221-1705
Plans, kits, lasers, power supplies

JDR Microdevices
1224 S. Bascom
San Jose, CA 95128
(408) 995-5430
Computers & parts

Jerryco Incorporated
607 Linden Place
Evanston, IL 60202
(312) 475-8440
All types surplus, batteries, gadgets

Marlin Jones Incorporated
P.O. Box 12685
Lake Park, FL 33403
(407) 848-8236

Maxim Integrated Circuits
120 N. San Gabriel
Sunnyvale, CA 94086
(408) 737-7600

Free A/D design guide,
A/D evaluation kit,
MAX 120 EV kit $50

Mondotronics Incorporated
1014 Morse Avenue, Ste. 11
Sunnyvale, CA 94089
(408) 734-9877
Biometal wire

Morse Security Group
12960 Bradley Avenue
Sylmar, CA 91392
Alarm controls

Mouser Electronics
2401 Hwy. 287 North
North Mansfield, TX 76063
(817) 483-4422
Low-cost parts, good selection

NAPCO Alarm Systems
333 Bayview
Amityville, NY 11701
(516) 842-9400
Alarm controls

Opto Electronics Inc.
5821 N.W. 14th Ave.
Ft. Lauderdale, FL 3334
Frequency counters

PAN - Com
P.O. Box 130
Paradise, CA 95967
(916) 534-0417
Kits, parts, plans

PMC Electronics
P.O. Box 11148
Marina del Rey, CA 90292
(213) 827-1852
Videophone kits

Prairie Digital
846 Seventeenth St.
Industrial Park
Prairie du Sac, WI 53578
Inexpensive A/D cards

Ramsey Electronics
793 Canning Parkway
Victor, NY 14564
(716) 924-4560
Radio kits

R&D Electronics
1224 Prospect Avenue
Cleveland, OH 44115
(216) 621-1121
Parts, surplus, power supplies

Real Time Devices
820 N. University Drive
State College, PA 16804
(814) 234-8087
Computer cards, I/O
accessories

Summit Electronics
(800) 678-6648
New electronic parts, good
selection

Tapeswitch of America
100 Schmitt Rd.
Farmingdale, NY 11735
(516) 694-6312
Tape switch sensors and mats

Teiresias Inc.
Dept. TR
Sag Harbor Turnpike
Sag Harbor, NY 11963
Floor-stress sensors

Thermo Disk Incorporated
1320 S. Main Street
Mansfield, OH 44907
(419) 756-5911
Thermal switches

Toko America
1250 Feehanville Dr.
Mount Prospect, IL 60056
(312) 297-0070
rf coils

Victory Engineering Co.
P.O. Box 559
Springfield, NJ 07081
Toko coils

World Magnetics
810 Hastings St.
Traverse City, MI 49684
Driveway low-pressure air switches

# Index

## A

ac-magnetic field detector, 25-27, **26**
accelerometer, 79-81, **80**
air switch, shallow-buried, 143
alcohol breath analyzer, 117-118, **117**, 122-123
amplifiers
  high-gain, 1-5, **2**
    light detector, 3, **4**
    metal detector, 2, **3**
    parts list, 5
    radiation detector, 4, **4**
    rf detector, 2-3, **4**
    telephone listening coil, 1-2, **3**
  op-amp amplifier
    differential, 69, **70**
    piezo-film sensors, 69, **69**
analog-to-digital interface, 129-130, **131**
auto burglar alarm, 181-183, **181**
auto immobilizer, 184-186, **185**
auto lock, antitheft, digital, 186-188, **187**
automatic emergency-lighting system, 190-191, **190**

## B

bar-graph displays
  analog pressure sensors bar-graph, 37, **38**
  auto-battery voltmeter, 183-184, **184**
battery voltmeter, auto battery, bar-graph display, 183-184, **184**
body-heat detector, 33-36, **34**, **35**, 140-141, **140**
breath analyzer, 117-118, **117**, 122-123
bridge circuits, 27-31, **28**, **29**, **30**
  dual variable dc, 28, **29**
  instrumentation-grade, 29-31, **30**
  Maxwell bridge, 31-33, **32**
  quad variable dc, 28-29, **30**
  single variable dc, 28, **29**
bugs, electronic-bug detection, 165-166
bus-timing diagram, multiplexed Hall-effect sensor, 99, **100**

## C

cameras, 147-150, **148**
  video motion detectors, 149
camp alarm, 226-230, **227**, **229**
capacitive proximity sensors, 207-209, **207**
carrier-current control systems, 211-216, **212**, **214**
Chang, 121
CHEMFETs, 121
chimney alarm, 235-240, **236**, **238**
circuits and systems, 169-222
  auto burglar alarm, 181-183, **181**
  auto immobilizer, 184-186, **185**
  bar-graph auto voltmeter, 183-184, **184**
  carrier-current control systems, 211-216, **212**, **214**
  digital antitheft auto lock, 186-188, **187**
  emergency-lighting system, 190-191, **190**
  high-performance alarm, 220-222, **220**
  home guard, 201-205, **202**
  latching alarm, 169-171, **170**
  location-display alarm system, 175-178, **176**
  motion detector, optical, 205-207, **206**
  motion detectors, microwave, 209-211, **210**
  multipurpose alarm system, 178-181, **179**
  optical motion detector, 205-207, **206**
  power-line fault detector, 188-190, **188**

circuits and systems (*cont.*)
  proximity sensor, capacitive, 207-209, **207**
  remote sensing, 171-173, **171**
  rf sniffer, 198-201, **199**
  siren, adjustable-rate, 191-193, **192**
  strobe flasher, 193-195, **193**
  telephone line monitor, 195-196, **195**
  telephone recorder, 196-198, **197**
  ultrasonic-sensor system, 217-220, **217**, **218**
  window/door alarm, 173-175, **173**, **174**
closed loop alarm systems, 153-154
computer interfacing, 125-132
  analog-to-digital interface, 129-130, **131**
  joystick interfaces, 125-127, **126**, **128**
  mouse interface, 128-129, **130**
  parts lists for all interfaces, 130-132
  serial interface, 127-128, **129**
  trackball interface, 128-129, **130**
control boxes, 159-160
  carrier-current control systems, 211-216, **212**, **214**
counters
  piezo-film sensors
    counter/counter interface, 69, **71**
current detector
  Hall-effect sensors, 20-22, **21**
  over/under current, 89-90, **90**
Currie, Jacque and Pierre, 63

## D

designing alarm systems, 151-167
  bugs, electronic-bug detection, 165-166
  closed or "supervised" loop alarm systems, 153-154
  control boxes, high-tech, 159-160

disguising the sensors, 153-154
doors, 151-153
fire reporting, 162-163
history of alarms, 156-158
installing alarm systems, 154-156
keys and locks, 151-153
lighting for crime prevention, 164-165
sirens, 160
strobe lights, 160
telephone dialers, 160
telephone dialers, digital, 164-164
wireless alarm systems, 160-162
wiring techniques, 154-156
diagnostic response, multiplexed Hall-effect sensor, 99
digital antitheft auto lock, 186-188, **187**
digital telephone dialers, 163-164
direction sensor, fiberoptic sensors, 45, **46**
disguising sensors in alarm systems, 153-154
door/window alarm, 173-175, **173**, **174**
doors, 151-153
DTMF alarm system, 240-247, **241**, **242**, **243**, **244**

## E

Earth-movement sensor, 59-61, **59**, **61**
electrolytic cells, 53, **53**
electronic level, 52-53, **52**
electroscope, 8-9, **9**
emergency-lighting system, 190-191, **190**
encoder
  optical rotary encoder, 47-50, **47**, **48**, **49**

## F

fence alarm sensors, 144
  laser, 145
fiberoptic sensing, 44-47, **45**

Index  281

bifurcated light pipe, 44, **44**, **45**
direction sensor, 45, **46**
parts list, 47
receiver, light-pipe, 45, **46**
field detector, ac-magnetic, 25-27, **26**
film sensors (*see* piezo-film sensors)
film-strip position sensor, 50-52, **50**, **51**, **52**
fire detectors, 145-147
  rate-of-rise detector, 145, **145**
fire reporting systems, 162-163
flasher circuit, strobe light, 193-195, **193**
flex switch, piezo-film sensors, 66, **66**
fluid detector, 87-89, **87**, **88**
force-sensing resistor (FSR), 72-77, **73**, **74**, **75**
  force-to-frequency converter, 76, **76**
  force-to-voltage converter, 75, **76**
  parts lists, 77
force-to-frequency converter, 76, **76**
force-to-voltage converter, 75, **76**

# G

gas sensing technology, 105-123
  alcohol breath analyzer, 117-118, **117**, 122-123
  back-cell type gas sensor, 119, **120**
  carbon-monoxide detectors, **110**, 111
  CHEMFETs, 121
  circuits using gas sensors, 108, **110**
  custom IC-TGS203 gas sensor, 111-112, **114**
  custom IC-TGS550 gas sensor, 113
  gas sensors, diagrams, **109**
  gas sensors, typical, **107**
  ISFETs, 121
  methane-gas detector, 115, **116**, 122
  multisensor arrays, 121-122
  oxygen sensor, 118-119, **118**
  ozone detectors, 121-122
  sensitivities of gas sensors, 111, **112**, **113**
  series 8 gas sensors, TGS813, 115
  setpoint variations, 115, **116**
  smoke detectors, 119, **120**
  solid-state gas sensors, **106-107**
  tin-dioxide sensors, operation, 107-108, **108**
  types of gases detected, 105
gaussmeter, 22-23, **22**
General Motors, 121
glass-breakage detector, 137, **138**

# H

Hall-effect sensors, 19-25
  applications for Hall-effect sensors, 23-24, **23**
  current-sensor, 20-22, **21**
  gaussmeter, 22-23, **22**
  metal detector, 20, **21**
  multiplexed, 98-103, **99**, **100**
    bus interface, 102, **102**
    bus-timing diagram, 99, **100**
    diagnostic response, 99
    diagram, 102, **103**
    signal response, 101
  parts lists for all sensors, 24-25
  switch, 19-20, **19**
  switch, ac, 20, **20**
heater, temperature-controlled, 16-17, **17**
high-performance alarm, 220-222, **220**
high-tech security projects, 223-274
  camp alarm, 226-230, **227**, **229**
  chimney alarm, 235-240, **236**, **238**
  DTMF alarm system, 240-247, **241**, **242**, **243**, **244**
  piezo vibration alarm, 223-226, **224**, **225**
  portable alarm, 247-257, 248,

high-tech security projects (*cont.*) 249, 250, 251, 252, 253, 254
  pyroelectric sensor, 230-235, **231**, **233**, **234**
  storm warning, 258-264, **259**, **260**, **261**
  video sentry, 264-274, **265**, **267**, **268**, **269**
history of alarms, 156-158
holdup switch security systems, 135-136
home guard, 201-205, **202**
humidity sensor, 42-44, **42**, **43**

## I

immobilizer, automotive, 184-186, **185**
infrared sensors (*see also* pyroelectric detectors), 140-141, **140**
  telescopes, passive IF, 144
installing alarm systems, 154-156
interfaces (*see* computer interfacing)
interior "space" detectors, 138
ionizing smoke detectors, 146
ISFETs, 121

## J

joystick interfaces, 125-127, **126**, **128**

## K

keys and locks, 151-153
Kynar, 63-64, **64**

## L

laser fence alarm sensors, 145
latching alarm, 169-171, **170**
level detector
  electronic level, 52-53, **52**
  piezo-film sensors, 69, **70**
light detector, 3, **4**, 10, **10**
light tube, 11, **11**
light-beam sensors, 138-139, **139**
light-level detector, 11-12, **12**
light-wave receiver, 51, **52**
light/dark switch, 9-13, **10**, **11**, **12**

lighting for crime prevention, 164-165
location-display alarm system, 175-178, **176**
locks and keys, 151-153
LVDT tiltmeters, 54, **54**, **55**

## M

magnetic switch security systems, 134-135, **135**
magnetic switch, piezo-film sensors, 67, **68**
magnetic tiltmeters, 54-55, **55**, **56**
Maxwell bridge, 31-33, **32**
metal detector, 2, **3**
  Hall-effect sensors, 20, **21**
  proximity detector, 83-84, **84**, 143
methane-gas detector, 115, **116**, 122
microwave alarm sensors, 141, **141**
microwave motion detectors, 209-211, **210**
motion detectors, 136-145
  air switch, shallow-buried, 143
  fence alarm sensors, 144
    laser, 145
  glass-breakage detector, 137, **138**
  interior "space" detectors, 138
  light-beam sensors, 138-139, **139**
  microwave, 141, **141**, 209-211, **210**
  optical, 205-207, **206**
  proximity sensors, 141-143, **142**
    capacitive, 142-143, **142**
    metal proximity, 143
  pyroelectric sensors, 140-141, **140**
  road-switch, magnetic, 143-144
  sound detector, 137, **138**
  telescopes, passive infrared, 144
  touch switch, capacitive, 143

Index  *283*

ultrasonic detectors, 139-140, **139**
vibration detectors, 136-137, **137**
video motion detectors, 149
mouse interface, 128-129, **130**
multiplexed Hall-effect sensor, 98-103, **99**, **100**
  bus interface, 102, **102**
  bus-timing diagram, 99, **100**
  diagnostic response, 99
  diagram, 102, **103**
  signal response, 101
multipurpose alarm system, 178-181, **179**

## O

op-amp amplifier, piezo-film sensors, 69, **69**
  piezo-film sensors, differential, 69, **70**
optical motion detector, 205-207, **206**
optical position light-wave receiver, 51, **52**
optical rotary encoder, 47-50, **47**, **48**, **49**
optical switch, 51, **51**
optical transceiver, 81-83, **82**
overcurrent detector, 89-90, **90**
oxygen sensor, 118-119, **118**

## P

passive infrared telescopes, 144
Penwalt Corporation, 63
photoelectric smoke detectors, 146-147, **146**
piezo-film sensors, 63-77
  accelerometer, 79-81, **80**
  applications for piezo-film sensors, 65
  counter/counter interface, 69, **71**
  Currie, Jacque and Pierre, piezoelectricity, 63
  flex switch, 66, **66**
  force-sensing resistor (FSR), 72-77, **73**, **74**, **75**
    force-to-frequency converter, 76
    force-to-voltage converter, 75
    parts lists, 77
  IR-sensing piezo-film, 65, **65**, **66**
  Kynar film, 63-64, **64**
  level detector, 69, **70**
  magnetic switch, 67, **68**
  op-amp amplifier, 69, **69**
  op-amp amplifier, differential, 69, **70**
  Penwalt Corporation, 63
  piezoelectricity, 63
  pressure detector with audio output, 69, **71**
  process of making piezo-film, 64-65, **64**
  PVDF film, 63-64, **64**
  pyroelectric detector, 65, **65**
  range of detection, 65
  singing switch, 67-69, **68**
  snap switch, 67, **67**
  thickness and configurations of piezo-film sensors, 65-66, **66**
  vibration alarm system, 223-226, **224**, **225**
piezoelectricity, 63
portable alarm, 247-257, **248**, **249**, **250**, **251**, **252**, **253**, **254**
position-sensitive detector (PSD), 55-57, **56**, **57**, 94-96, **94**, **95**, **96**
  film-strip position, 50-52, **50**, **51**, **52**
power-line fault detector, 188-190, **188**
pressure sensors, 36-40, **36**
  analog bar-graph, 37, **38**
  computer interface, 37-38, **39**
  parts lists for all sensors, 38-40
  piezo-film sensors, audio output, 69, **71**
  switch, 36-37, **37**
proximity detector, 83-84, **84**, 141-143, **142**
  capacitive, 142-143, **142**, 207-209, **207**
  metal proximity, 143

PVDF film, 63-64, **64**
pyroelectric detector, 33-36, **34**, **35**, 65, **65**, 140-141, **140**, 230-235, **231**, **233**, **234**

# R

radiation detector, 4, **4**
rate-of-rise heat detector, 145, **145**
receivers
  light-pipe (fiberoptic) receiver, 45, **46**
  light-wave receiver, 51, **52**
recorder, telephone recorder, 196-198, **197**
remote sensing system, 171-173, **171**
rf detector, 2-3, **4**, 198-201, **199**
road-switch, magnetic, 143-144
rotary encoder, optical, 47-50, **47**, **48**, **49**

# S

sensor mat security systems, 136
sensors and detection circuits, 1-61
  ac-magnetic field detector, 25-27, **26**
  amplifier, high-gain, 1-5, **2**
  bridge sensors, 27-31, **28**, **29**, **30**
    dual variable dc, 28, **29**
    instrumentation-grade, 29-31, **30**
    Maxwell bridge, 31-33, **32**
    quad variable dc, 28-29, **30**
    single variable dc, 28, **29**
  Earth-movement sensor, 59-61, **61**
  electroscope, 8-9, **9**
  fiberoptic sensing, 44-47, **45**
    bifurcated light pipe, 44, **44**, **45**
    direction sensor, 45, **46**
    parts list, 47
    receiver, light-pipe, 45, **46**
  film-strip position sensor, 50-52, **50**, **51**, **52**
  Hall-effect sensors, 19-25
    applications, 23-24, **23**
    current-sensor, 20-22, **21**
    gaussmeter, 22-23, **22**
    metal detector, 20, **21**
    parts lists, 24-25
    switch, 19-20, **19**
    switch, ac, 20, **20**
  humidity sensor, 42-44, **42**, **43**
  light detector, 3, **4**, 10, **10**
  light tube, 11, **11**
  light-level detector, 11-12, **12**
  light/dark switch, 9-13, **10**, **11**, **12**
  Maxwell bridge, 31-33, **32**
  metal detector, 2, **3**
  optical rotary encoder, 47-50, **47**, **48**, **49**
  pressure sensor, 36-40, **36**
    analog bar-graph, 37, **38**
    computer interface, 37-38, **39**
    parts lists, 38-40
    switch, 36-37, **37**
  pyroelectric detector, 33-36, **34**, **35**
  radiation detector, 4, **4**
  receiver, light-wave, 51, **52**
  rf detector, 2-3, **4**
  static electricity detector, 7, **7**
  switches, optical, 51, **51**
  telephone listening coil, 1-2, **3**
  temperature sensors, 13-19
    adjustable, 16, **17**
    analog, 14, **14**
    bonding sensor to source, 15-16
    computer interface, 16
    extended-range, 14, **15**
    heater, temp-controlled, 16-17, **17**
    parts lists, 17-19
    remote, 14, **15**
    thermistors, 13, **13**
    thermocouples, 13-14, **14**
  tiltmeters, 52-58
    electrolytic cells, 53, **53**
    electronic level, 52-53, **52**
    LVDT, 54, **54**, **55**
    magnetic, 54-55, **55**, **56**
    parts lists, 57-58
    position-sensitive detector

(PSD), 55-57
touch switch, 5-7, **5**, **6**
toxic-gas sensor, 40-42, **41**
serial interface, 127-128, **129**
shallow-buried air switch, 143
signal response, multiplexed Hall-effect sensor, 101
singing switch, piezo-film sensors, 67-69, **68**
sirens, 160
   adjustable-rate, 191-193, **192**
smoke detector, 85-86, **86**, 145-147
   gas sensing technology, 119, **120**
   ionizing smoke detectors, 146
   photoelectric type, 146-147, **146**
   rate-of-rise heat detector, 145, **145**
snap switch, piezo-film sensors, 67, **67**
sound detector, 137, **138**
space detectors, 138
speed detectors (*see* tachometers)
static electricity detector, 7, **7**
storm warning, 258-264, **259**, **260**, **261**
strobe lights, 160
   flasher circuit, 193-195, **193**
"supervised" loop alarm systems, 153-154
suppliers, 275-278
switches
  air switch, shallow-buried, 143
  flex switch, piezo-film sensors, 66, **66**
  fluid detector, 87-88, **87**
  force-sensing resistor (FSR), piezo-film sensors, 72-77, **73**, **74**, **75**
  holdup switches in security systems, 135-136
  light/dark switch, 9-13, **10**, **11**, **12**
  magnetic switch, piezo-film sensors, 67, **68**
  magnetic switches in security systems, 134-135, **135**
  optical switch, 51, **51**
  pressure sensors, 36-37, **37**
  road switch, magnetic, 143-144
  singing switch, piezo-film sensors, 67-69, **68**
  snap switch, piezo-film sensors, 67, **67**
  tape-switch mats in security systems, 136
  touch switch, 5-7, **5**, **6**
    capacitive, 143

# T

tachometer, 91-94, **92**
   capacitive tachometers, 91, **93**
   computer interface, 91-92, **93**
   filtered-input tachometers, 91, **92**
   parts lists, 93-94
tape-switch mat security systems, 136
telephone dialers, 160
   digital, 163-164
telephone line monitor, 195-196, **195**
telephone listening coil, 1-2, **3**
telephone recorder, 196-198, **197**
telescopes, passive infrared, 144
temperature sensors, 13-19
   adjustable, 16, **17**
   analog, 14, **14**
   bonding sensor to source, 15-16
   computer interface, 16
   extended-range, 14, **15**
   heater, temperature-controlled, 16-17, **17**
   parts lists for all sensors, 17-19
   remote, 14, **15**
   thermistors, 13, **13**
   thermocouples, 13-14, **14**
thermistors, 13, **13**
thermocouples, 13-14, **14**
tiltmeters, 52-58
   electrolytic cells, 53, **53**
   electronic level, 52-53, **52**

tiltmeters (*cont.*)
  LVDT, 54, **54**, **55**
  magnetic, 54-55, **55**, **56**
  parts lists, 57-58
  position-sensitive detector (PSD), 55-57, **56**, **57**
touch switch, 5-7, **5**, **6**
  capacitive, 143
toxic-gas sensor, 40-42, **41**
trackball interface, 128-129, **130**
transceiver, optical, 81-83, **82**
twilight sensor, 97-98, **97**, **98**

## U

ultrasonic detectors, 139-140, **139**
ultrasonic-sensor alarm system, 217-220, **217**, **218**
undercurrent detector, 89-90, **90**

## V

vibration detectors, 136-137, **137**
  piezo vibration alarm, 223-226, **224**, **225**
video motion detectors, 149
video sentry, 264-274, **265**, **267**, **268**, **269**
videophone, 103-105, **104**
voltmeter, auto battery bar-graph voltmeter, 183-184, **184**

## W

window foil security systems, 133-134, **134**
window/door alarm, 173-175, **173**, **174**
wireless alarm systems, 160-162
wiring techniques for alarm systems, 154-156

# Other Bestsellers of Related Interest

**Build Your Own Home Security System**
—*Delton T. Horn*
Now, even if you only have a basic knowledge of electronics, you can protect the safety and value of your property without buying an expensive security system. Horn shows you how to build and install fire and smoke alarms, gas and temperature range detectors, motion detectors, power shut-down/reset timers, a microwave meter and many more projects that require only low-cost and easy-to-find components and a few basic tools to construct. Seven different burglar alarms for home and auto make this book particularly valuable.
**0-07-030393-2     $17.95 Paper**

**Electronic Alarm and Security Systems: A Technician's Guide**
—*Delton T. Horn*
Here is a complete guide for professionals and electronics enthusiasts who want to learn how to install, maintain, and repair all kinds of alarm and security systems. Drawing on his own experience as an alarm system technician, Horn provides you with step-by-step guidance, including several chapters devoted to intermediate- and advanced-level troubleshooting. Other chapters discuss basic alarm system elements, planning the most cost-efficient systems for the job, and master control centers—including alarm location indicators and emergency bypass systems.
**0-07-030528-5     $39.95 Hard**
**0-07-030529-3     $34.95 Paper**

**Encyclopedia of Electronic Circuits, Volume 5**
—*Rudolf F. Graf and William Sheets*
This giant collection of original circuits, organized alphabetically into more than 100 categories, includes an extensive index to all the circuits presented, as well as for those found in each of the four previous editions. For each circuit, Graf and Sheets provide you with source references, component information, and other information regarding adjustments or alignments as needed. You will find diagrams and schematics for circuits to be used in: alarm and security systems; smoke, moisture, and metal detectors; amplifiers, receivers, and transmitters; and more.
**0-07-011076-X     $60.00 Hard**
**0-07-011077-8     $34.95 Paper**

### How to Test Almost Everything Electronic, 3rd Edition
—*Delton T. Horn*
This bestselling guide should be the first word for you if you are a beginner level hobbyist or student who wants to understand and use today's electronic test equipment. Light on theory and mathematical calculations, it's a practical handbook and reference that clearly explains how common testing devices are used to pin-point problems in everything from TV sets and computers to automotive electrical systems. It has been revised to include current testing techniques and new chapters on mechanical repairs and flowcharting.
**0-07-030406-8     $15.95 Paper**

### Electronic Projects to Control Your Home Environment
—*Delton T. Horn*
This book contains more than 25 projects—with easy-to-follow instructions, schematics, and parts lists—for the hobbyist or do-it-yourselfer who wants to make low-cost sensing and detection equipment. At low cost, you can build electronic devices that accurately identify, measure, and monitor environmental risks in the home. Projects include an electronic and digital thermometer, humidity and temperature alerts, a heat leak locator, visual and audible flooding alarms, an air ionizer, microwave and radiation monitors, and more.
**0-07-030416-5     $26.95 Hard**
**0-07-030417-3     $16.95 Paper**

### How to Build Earthquake, Weather, and Solar Flare Monitors
—*Gary G. Giusti*
This book shows how to build inexpensive, yet highly sensitive and accurate devices for monitoring seismic activity, meteorologic phenomena such as lightning strikes, and solar flares. Ideal for electronics and science hobbyists, inventors, and others interested in weather, it includes more than 40 projects, which can be combined to construct a complete scientific monitoring station. The author takes you from basic to more advanced projects, including a seismometer, earthquake alarm, geiger counter, and more.
**0-07-025209-2     $19.95 Paper**

## How to Order

**Call 1-800-822-8158**
24 hours a day,
7 days a week
in U.S. and Canada

**Mail this coupon to:**
McGraw-Hill, Inc.
P.O. Box 182067
Columbus, OH 43218-2607

**Fax your order to:**
614-759-3644

**EMAIL**
70007.1531@COMPUSERVE.COM
COMPUSERVE: GO MH

### Shipping and Handling Charges

| Order Amount | Within U.S. | Outside U.S. |
|---|---|---|
| Less than $15 | $3.50 | $5.50 |
| $15.00 - $24.99 | $4.00 | $6.00 |
| $25.00 - $49.99 | $5.00 | $7.00 |
| $50.00 - $74.49 | $6.00 | $8.00 |
| $75.00 - and up | $7.00 | $9.00 |

# EASY ORDER FORM—
# SATISFACTION GUARANTEED

Ship to:
Name _____
Address _____
City/State/Zip _____
Daytime Telephone No. _____

*Thank you for your order!*

| ITEM NO. | QUANTITY | AMT. |
|---|---|---|
|  |  |  |
|  |  |  |

Method of Payment:
☐ Check or money order enclosed (payable to McGraw-Hill)

| | |
|---|---|
| Shipping & Handling charge from chart below | |
| Subtotal | |
| Please add applicable state & local sales tax | |
| TOTAL | |

☐ DISCOVER  ☐ AMERICAN EXPRESS Cards
☐ VISA  ☐ MasterCard

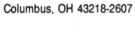

Account No. ☐☐☐☐☐☐☐☐☐☐☐☐☐☐☐☐

Signature _____ Exp. Date _____
*Order invalid without signature*

**In a hurry? Call 1-800-822-8158 anytime,
day or night, or visit your local bookstore.**

Key = BC95ZZA

# About the Author

Thomas Petruzzellis is a professional electronics technician who is employed at the State University of New York at Binghamton. He is responsible for designing, building and maintaining many different types of electronic devices involving sensors.